"*Shamans Among Us* is a thought-provoking essay on the idea that the psychology of people who we today diagnose with schizophrenia is akin to that of shamans in ancestral societies. Joseph Polimeni's scholarly book challenges several traditional concepts of both evolutionary biology and medicine. I strongly recommend it to all those who dare to think outside the box."

– Martin Brüne, MD, author of *Textbook of Evolutionary Psychiatry: The Origins of Psychopathology*

"What an aptly named book. Joseph Polimeni raises the interesting notion that present-day schizophrenics are, in a real sense, descendants of latter-day shamans — evolutionary selectees who served the important societal function of communication with the supernatural."

– David Koulack, PhD, author of *To Catch a Dream: Explorations of Dreaming*

"As with other evolutionary theories, it casts a new light on matters which were previously confusing and even inexplicable."

– John Price, DM, MRCP, MRCPsych, DPM, co-author of *Evolutionary Psychiatry: A New Beginning* and *Prophets, Cults and Madness*

Shamans Among Us

Schizophrenia, Shamanism and the Evolutionary Origins of Religion

Joseph Polimeni, MD FRCPC

John Scott Price, DM, MRCP, MRCPsych, DPM, is retired from psychiatric practice in the UK National Health Service. Previously he worked for the Medical Research Council, in the Psychiatric Genetics Research Unit and in the Clinical Research Centre. For many years he was European Editor of the ASCAP (Across Species Comparisons and Psychopathology) Newsletter. He is co-author with Anthony Stevens of *Evolutionary Psychiatry* (Routledge 1996, 2000) and *Prophets, Cults and Madness* (Duckworth, 2000). He is interested in conflict and reconciliation, especially as depicted in literature, such as in the Indian epic story *The Mahabharata*. He is currently Co-chair of the Section on Evolutionary Psychiatry of the World Psychiatric Association, and was recently Chair of the Section on Psychotherapy of that organization. Visit www.johnprice.me.uk.

©2012 Joseph Polimeni
All rights reserved

No part of this book may be used or reproduced in any manner without written permission from the author, except in the context of reviews.

Errors or omissions will be corrected in subsequent editions.

Product and company names mentioned herein may be the trademarks of their respective owners.

Publisher's Cataloging-in-Publication data

```
Polimeni, Joseph.
    Shamans among us : schizophrenia, shamanism and the
 evolutionary origins of religion / Joseph Polimeni, MD FRCPC.
     p. cm.
    Includes bibliographical references and index.
    ISBN 978-1-300-43091-9 (pbk.)
    ISBN 978-1-300-45851-7 (Hardcover)
    ISBN 978-1-300-47446-3 (epub)
    ISBN 978-1-300-47444-9 (PDF)

    1. Schizophrenia --Religious aspects. 2. Shamanism. 3. Human
 evolution --Religious aspects.  4. Psychology, Religious. 5.
 Religion and Psychology. 6. Psychiatry and religion. I. Title

 BL53 .P645 2012
 128/.2 -dc23
```

Revision Date: December 5, 2012

Shamans Among Us was edited, designed and typeset by
Amy Brown (http://www.amyrbrown.ca/).

Photo of the author by Adrian Polimeni. Cover photo by the author.

Tables and figures prepared by John Funk. Thanks to Nancy Wells and Aileen Hutchison for their help creating this book.

To Adrian and Julian

*Our greatest responsibility
is to be good ancestors.*
— Jonas Salk

Contents

Foreword by John Price		ix
Acknowledgments		xiii
1	Evolution Changes Everything	1
2	The Sculpting of Schizophrenia	15
3	The Schizophrenia Paradox	29
4	Is Schizophrenia a Disease?	49
5	The Silver Lining of Psychosis	75
6	All Things Evolution	99
7	The Evolutionary Origins of Religion	135
8	Shamanism	151
9	On the Edge of Insanity	169
10	Contemporary Delusions and Hallucinations	191
11	Finishing Touches	209
References		223
Index		265
Colophon		272

Foreword

This book is about man's contact with the supernatural, and about the people who make that contact, and how, if their contact does not go well, they run the risk of being labelled as mad. The author, as an atheist, analyses this process according to the latest knowledge of anthropology and psychiatry.

Contact with the supernatural has at least three advantages for a group of human beings, and these advantages may be described as morals, divination and morale. There is little doubt that a group with a good moral code functions well, and better than groups lacking a moral code. If the code has the backing of the supernatural, it is more likely to be followed by the group members; and if violation of the moral code is punishable by supernatural forces rather than by other group members, then so much better for the peace of the group.

Divination is important for deciding between alternatives. Differing and mutually incompatible courses of action are likely to present themselves to a group. Should they plant one crop or another, should they go to war, and if so with whom? It is likely that members of the group will have different opinions, and therefore endanger the group with conflict; but if the will of the gods is ascertained by divination, then this divine will overrides those contrary wills of the group members and peace within the group is maintained.

When the group goes to war, they are likely to fight harder if they know that the gods are on their side, so whoever is able to reassure them on this point is in a position to raise group morale.

These three advantages are dependent on contact with the gods being restricted to a single individual. If everyone can ascertain the will of the gods, then the situation is no better than before: each group member may ascertain a different moral code from the gods, so that there is no uniformity of behaviour, and transgressions of the code will not be punished. If every group member is skilled in divination, then there are likely to be as many predictions from the gods as there are group members, so that there will be no unity of action. In the case of warfare, all members of the group are likely to divine that the gods are on their side, but each may be praying to a different god, so that even here the unity of group purpose is not likely to be maintained.

How does the group identify the member who has this special conduit to the supernatural, and so confer on him the role of shaman? One qualification may be that he (or she) has been to places that other group members have not visited, such as other planets or the centre of the earth. Another is that the shaman hears voices when no one is speaking, and these voices are therefore likely to be the voices of supernatural entities. Also, he makes his pronouncements with a sense of conviction which is beyond the certainty of ordinary group members. Thus the shaman shares with the psychotic patient the experience of astral travel, the receipt of auditory hallucinations and the utter conviction of the deluded. We know that a significant proportion of children have auditory hallucinations, but that these tend to disappear with age; but if a replacement shaman is needed for the group, one of these children may be encouraged to continue them, and so provide an assistant for the shaman, and eventually a replacement. If no trainee shaman is required, those young people who continue to have hallucinations are likely to be regarded as psychotic.

Thus it seems likely that the whole human race is genetically programmed to believe in the supernatural, and that a very small proportion of us are genetically programmed to communicate with the supernatural, and if the epigenetic pathway is favourable, these latter end up as shamans or other forms of magico-religious practitioner, whereas if the epigenetic pathway is unfavourable they end up as psychiatric patients. Joseph Polimeni argues the case for this scenario with scholarship and clarity.

Inevitably, this book will attract much negative criticism. It will be attacked from many quarters. The anti-evolutionary psychology lobby will claim it is unscientific and merely a collection of just-so stories. The "medical model" psychiatrists will attack it because it threatens their perception of themselves as medical men. The anti-group selection lobby will deride it as "naive group selectionism." Some politically motivated people will damn it because it does not conform to the tenets of Marxism-Leninism. The pharmaceutical industry will dislike it because it suggests ways of dealing with psychosis other than drugs. The managed-care people will hate it because it will take more doctor time than the mere prescription of drugs. And, needless to say, religious people will disapprove of it because it offers a scientific explanation of spirituality. However, as with other evolutionary theories, it casts a new light on matters which were previously confusing and even inexplicable. As Theodosius Dobzhansky entitled one of his papers, "Nothing in biology makes sense except in the light of evolution" (Dobzhansky 1973).This applies also to human behaviour, in its normal and its pathological forms.

Joseph Polimeni's book makes a case for the shaman as a facilitator of group processes and an agent of cohesion for the group. It may seem odd that I am strongly recommending this book when I co-authored *Prophets, Cults and Madness* with Anthony Stevens (Duckworth, 2000) in which we made a case for the medico-religious practitioner being an agent of group splitting, as the prophet proclaimed a new world order which made him and his followers unacceptable to the group, as a result of which they went off to a "promised land." Is the medico-religious practitioner an agent of homeostasis or an agent of change? It seems likely that he is both. While the group is growing in size he is an agent of homeostasis, but when the group gets to a certain size, he uses the same talents which made him a shaman to become a prophet of a New Religious Movement and so split the group. Like an amoeba, the same genetic material which makes the group cohere changes function when it is time to split, and then it organises meiosis and enables the amoeba to divide into two. Rapid group splitting is important in evolution because it favours selection at the group level rather than the individual level, enables the members of the group to put the group interest before their own private interests, and so makes the group more effective in its competition with other groups (Price 2010). The idea of selection at the level of the group as well as at the level of the individual and the gene has been out of fashion for half a century, but is coming back into mainstream biology as a result of arguments made by David Sloan Wilson and Edward O. Wilson (Wilson 1997, Wilson and Sober 1994, Wilson 2012). The human race is eusocial, like ants and termites and some bees and wasps, and this has enabled it to be almost as successful in conquering the earth as the eusocial insects.

— John Price, DM, MRCP, MRCPsych, DPM

Acknowledgments

Schizophrenia may be the most enigmatic medical condition confronting twenty-first-century physicians. Incredibly, about fifty million people worldwide are afflicted by it. It is not easy living with schizophrenia; I know this both as a psychiatrist and as the nephew of a man stricken with the condition. The daily struggle of living with schizophrenia is an enormously important matter, but this is not a book about personal ordeals. This book is about a scientific theory, and as such will sometimes seem detached and impersonal. I do, however, wish to express my gratitude to the many patients who have helped me better understand psychiatric conditions. Some patients provide narratives that are so insightful and intellectual in spirit that they make these unfamiliar experiences a little more understandable to the rest of us. I am also indebted to the many patients who, through the years, volunteered their time for research studies.

It is not easy to find colleagues who are on the same wavelength, especially when one is challenging the core axioms of one's field. I was fortunate to partner with Dr. Jeff Reiss when I began formulating new arguments for the shamanism theory of schizophrenia. Jeff is an excellent clinician with a broad knowledge of psychiatric literature. He has two qualities that are rarely found together in the same person: imagination and a keen mind for detail. Both assets are invaluable when formulating novel evolutionary theories. Without him, the original articles on shamanism would have been less authoritative and less fun to do.

Since the late nineteenth century, eminent psychiatrists such as Karl Kahlbaum and Sigmund Freud understood that evolution could apply to mental disorders. In 1964, Julian Huxley, Ernst Mayr, Humphrey Osmond and Abram Hoffer were the first to consider evolutionary theory for a specific psychiatry problem. Then, in 1967, John Price went one step further: he authored the first comprehensive theory explaining the raison d'être of a mental disorder, depression, in a paper entitled "The Dominance Hierarchy and the Evolution of Mental Illness." John is still full of new ideas and no one has thought about evolutionary psychiatry longer than he has — someday John Price will be recognized as the father of evolutionary psychiatry. Ten years ago, when we published a hypothesis that seemed to oppose his own theory of schizophrenia, John nevertheless complimented us on our initial effort. He encouraged us to pursue these evolutionary questions further. I am thankful he did. I later discovered that such honorable gestures are not so common in the competitive world of science. For me, this story has a happy ending: we have both since learned that our respective theories could very well be complementary.

I would also like to acknowledge those colleagues who have always been generous with their time and shared my enthusiasm for various research projects cited in the book: Dr. Daryl Gill, Dr. Jitender Sareen, Dr. Darren Campbell, Brendon Foot and Breanna Sawatsky. I would also like to thank Dr. Don Head and Cheryl Maxom — health care professionals who are not researchers, but are good communicators. They read over earlier drafts to help make an academic book readable to the general audience.

On the day I spent perusing old medical records at the Selkirk Mental Health Centre, Janice Farion and her staff were extremely accommodating to a stranger with a strange idea.

I am most grateful to my editor, Amy Brown, who helped make the book readable to both academics and non-experts. Right off the bat, she completely understood my vision for the book and made it better. Also, a special thanks to Kalman Glantz and Gary Bernhard at EvoEbooks.

Last, my spouse, Christine. She is an experienced family physician with firsthand knowledge of every common medical disease. I was lucky to have an in-house consultant while investigating the role of evolution in medical disorders. Writing a book is a self-centered endeavor — it takes a lot of time and disrupts the cadence of family schedules. Christine has always been supportive and selflessly encouraging; she has made it easy for this psychiatrist to allay most of his guilt.

– 1 –
Evolution Changes Everything

> "It's unbelievable how much you don't know about the game you've been playing all your life."
> — Mickey Mantle

This is the story of an idea casually introduced a little over two centuries ago in the remote regions of Siberia. In the late eighteenth century a number of Russian scientific expeditions began exploring the Siberian frontier. A young research assistant, Vasilii Zuev, became one of the first Westerners to witness the religious rituals of the frontier tribes. Amazed by the eccentricities of their shamans, Zuev surmised that such behaviors were a medical form of craziness (Znamenski 2007). For many years this eccentric notion lingered, neither taken very seriously nor completely disproven.

In its simplest form, the shamanistic theory of schizophrenia says that people with schizophrenia are the modern manifestation of prehistoric tribal shamans. In other words, the inborn cognitive factors or personality style that would have predisposed certain people to become shamans is the same psychological mindset that underlies schizophrenia. Although the idea is straightforward, a proper evaluation of the hypothesis is complicated and must take into account the latest discoveries from psychiatry, medical history, evolutionary science, anthropology, psychology, religious studies and genetics. The primary purpose of this book is to put forward a contemporary version of the shamanistic theory of schizophrenia.

In hunting and gathering societies, shamans are designated experts who heal the sick, institute magical curses, carry out divination rituals and lead religious ceremonies. In older literature shamans were often referred to as sorcerers, medicine men, diviners, exorcists or witch doctors. Since the late 1800s, every generation of anthropologists has had an advocate who points out behaviors common to both shamans and the insane. Such resemblances can be interesting, even compelling; however, they are only a first step towards an integrative theory.

In 2002, Dr. Jeff Reiss and I published a brief paper in the journal *Medical Hypotheses* entitled "How shamanism and group selection may reveal the origins of schizophrenia" (Polimeni and Reiss 2002). This book refines and extends our introductory arguments on the topic. It is my intention to elevate centuries-old ideas about shamanism and psychosis to the level of systematic theory.

Over forty years have elapsed since the last psychiatrist expanded on the possible connection between shamanism and schizophrenia. In 1967, Julian Silverman built on the ideas of his predecessors in a brief paper entitled "Shamans and Acute Schizophrenia." In his introduction Silverman remarked, "the principal advantages of this formulation as compared to previous formulations of this problem lie in the clinical and experimental evidence available to support it and in its greater cross-cultural applicability" (Silverman 1967). It is precisely this sentiment that inspires the material in this book. Over the last few decades, more and more has been learned about schizophrenia and shamanism. These new discoveries allow us to assemble an even more comprehensive theory.

Schizophrenia is a serious mental condition characterized by a number of seemingly anomalous behaviors. Psychosis, which usually first appears in young adulthood, is its most dramatic symptom. Psychotic experiences can be subdivided into delusions, hallucinations and disjointed thoughts. Even before modern treatments came into existence, the natural course of psychosis tended to fluctuate — sometimes to the point of remission. Schizophrenia can be accompanied by subtler but more enduring symptoms such as poor social acumen, blunted affect (emotional state) and diminished motivation. In Western societies, occupational impairment is usually severe, as evidenced by unemployment rates approaching 90 percent (Marwaha and Johnson 2004; Perkins and Rinaldi 2002). This is just an abbreviated description of a very complex behavioral phenomenon. As our story unfolds, the reader will become familiar with schizophrenia and its most obvious symptom: psychosis.

Patients designated with schizophrenia tend to exhibit a cluster of characteristic symptoms and accordingly, there is reasonable *inter-rater reliability* (Tamminga et al. 2010, 13); in other words, physicians will usually agree on the diagnosis. However, inter-rater reliability does not necessarily translate to diagnostic validity. The concept of validity is contestable because it implies absolute truth behind an observation. The only way of having full confidence in the validity of any conclusion about a phenomenon is by understanding all of its relationships with the rest of nature — something that may never be completely achieved. Without a thorough understanding of the mechanics underlying an observable phenomenon, discussions about validity are almost moot. In the case of schizophrenia, many symptoms of the condition overlap with other identified disorders — that also have murky boundaries and are not yet well understood — such as bipolar disorder, schizoaffective disorder (characterized by symptoms of both schizophrenia and bipolar disorder) and delusional disorder. Schizophrenia also tends to blend with normality, as sub-threshold cases certainly exist (schizotypal or schizoid personality disorder, for example). Rather than endeavoring to contain schizophrenia inside a discrete category, some researchers have argued that psychosis lies on a spectrum seamlessly blending with normality and other psychotic disorders (dimensional approaches) (Esterberg and Compton 2009). The bottom line is that the precise inner workings and boundaries of schizophrenia are unknown.

Of all the serious psychiatric problems, schizophrenia is in some ways, the most mysterious. One perplexing attribute is how patients with schizophrenia can immerse themselves in normal and psychotic thinking simultaneously. I recall an emergency room encounter with a patient who was beginning to believe he was Jesus. I pressed for clarification; was he or was he not Jesus? There was a long pause and finally the patient replied that he wasn't sure. I persisted with this line of questioning: "If you are truly the Messiah, why would you seek psychiatric treatment?" He glossed over the query but did point out that he had been treated successfully with medications the last time he thought he was Jesus.

Although this clinical case happened many years ago, it has stood out for me as the first time I witnessed the precise moment when a mind teeters between normality and insanity. Given that this patient was intelligent and free of any other irrational thoughts, his indifference to such a gross lapse in logic was especially puzzling. (One of my old professors used to describe this fairly common psychotic phenomenon as "double bookkeeping.")

It would not be surprising to have a computer spew out nonsensical messages after falling off a desk. Likewise, it would not be profoundly bizarre to have somebody converse in scrambled jargon after getting bonked on the head. However, I have always found it bewildering how a tenuous sliver of insanity survives ensconced in an otherwise functioning brain. This kind of clinical case inevitably provokes a number of questions. How many neurons need to be sidetracked before a person slips into psychosis? How different is a typical brain from one that mistakenly believes he is Jesus? Is there a pattern or underlying purpose to psychosis? And finally, what do psychotic phenomena tell us, if anything, about our own bearings on reality?

Using Evolutionary Theory to Investigate a Biological Enigma

Schizophrenia is one of the greatest scientific enigmas. Despite a hundred years of study, there is still no cohesive understanding of the anatomical, physiological and psychological workings underlying schizophrenia. The very complexity of schizophrenia could be a sufficient explanation for its incomprehensibility. On the other hand, when a scientific problem is so intractable, one wonders about the inadvertent propagation of an unsuspected error.

One misplaced "fact" can prevent a scientific field from advancing towards a comprehensive theory for decades or more. In the late nineteenth century, for example, the implications of Maxwell's equations became increasingly bewildering to physicists. That is, until Albert Einstein challenged one obvious "fact": that time was immutable and constant. Einstein theorized that time was susceptible to change. Although extremely counterintuitive, the idea that time could vary reconciled a number of paradoxical phenomena in physics. In the back of my mind I have always wondered whether schizophrenia's elusiveness is simply due to one counterintuitive, misplaced "fact."

One way to investigate a puzzling medical condition is to trace its evolutionary history. No biologist would dispute that it's possible that evolutionary forces were involved in the genesis of some psychiatric conditions. (The evidence for evolution's role in psychiatric conditions will be reviewed in later chapters.) However, although almost every psychiatrist believes in the theory of evolution, very few readily acknowledge that evolutionary principles have any direct application to their field. There are several reasons for the dismissal of something as fundamental as evolutionary theory by the psychiatric establishment.

To understand evolution, it is helpful to frame it as the study of genes through time. The equation *genes + time = evolution*, clearly oversimplifies complex evolutionary processes but it will serve our purpose, which is to explain why evolution has been overlooked by the psychological sciences. For psychiatric conditions, their genetic underpinnings are essentially unknown and their history, beyond the last century, poorly documented. Thus, the two most important inputs of the evolution equation — genes and time — are blank. This makes the evolutionary study of psychiatric conditions difficult, but not impossible. The vast majority of psychiatrists have just not bothered thinking about the importance of evolutionary principles to psychiatric conditions, due to the dearth of reliable material supporting the field. The scarcity of requisite knowledge, however, in no way diminishes the potential importance of the matter. I cannot emphasize this point enough: schizophrenia may never be completely understood without an attempt to understand its inferable evolutionary path from genesis (perhaps, many thousands of years ago) to modern presentation in a twenty-first century emergency room. Throughout this book, I will illustrate new and indirect ways to uncover the possible evolutionary machinations behind psychiatric conditions.

There are other reasons why evolution has been relatively ignored in psychiatry. Psychiatrists have a longstanding tendency to overestimate environmental influences upon psychiatric conditions, to the relative exclusion of innate determinants. The eventual acceptance of nature over nurture for certain psychiatric ailments only occurred because overwhelming epidemiologic evidence was presented, not because this is the field's opening gambit. If one underestimates the role of genetics in psychiatry, then it follows that evolution must take a backseat.

The second reason for the marginalization of evolution in psychiatry is the near-absence of formal training in evolutionary science among medical school graduates. Although medical doctors, including psychiatrists, may possess some familiarity with the general principles of evolution, a refined understanding of evolution's possible role in medical conditions requires more than an introductory course in evolution. At this time, there may be only a dozen or so psychiatrists with more than a passing interest in the field. Accordingly, evolutionary perspectives represent a limited voice in the psychiatric literature.

The third major reason for the bypassing of evolutionary explanations is based on a not-unreasonable notion: that psychiatric conditions are just disadvantageous byproducts of more important evolutionary processes. For example, spinal disc herniation is a fairly common medical ailment, especially problematic to hominid species (compared to other mammals) due to our upright posture. At some point in our evolutionary history, the advantages of upright posture must have more-than compensated for all the potential downsides, like herniated discs; otherwise upright posture would not have become the norm. In other words, it is upright posture and locomotion that primarily dictate their own evolutionary path and not the smaller associated glitches. Spinal disc herniation is a secondary issue — something that evolution tolerates. Similarly, psychiatric conditions like schizophrenia are generally viewed as glitches that can occur in super-intelligent, complex brains. At first glance, this notion seems reasonable and may, in fact, be an appropriate account for some kinds of emotional problems. However, there is growing evidence that this perspective is not applicable to schizophrenia.

On closer examination, it becomes evident that schizophrenia does not share the typical attributes of an evolutionary afterthought. I will delve into this issue in greater depth in subsequent chapters; for now it is only important to know that this issue has been widely acknowledged by evolutionary psychiatrists and is known as the "schizophrenia paradox" (Brüne 2008). To put it simply, a major component of our contemporary formulation of schizophrenia seems to be at odds with evolutionary theory. If a psychiatric "fact" is at odds with a fundamental law of evolution, it is that psychiatric detail that is likely wrong. Barkow, Cosmides and Tooby are pioneers of the field of evolutionary psychology and in their classic book *The Adapted Mind* they emphasized this very point (1992, 4):

"... to propose a psychological concept that is incompatible with evolutionary biology is as problematic as proposing a chemical reaction that violates the laws of physics." The schizophrenia paradox has been cast aside because psychiatrists are rarely interested in evolutionary science. It is, however, misguided to neglect this "inconvenient truth," because investigating schizophrenia's place in evolutionary history has the potential to be enormously elucidating.

A number of possible solutions to the schizophrenia paradox have been offered but none have gained universal acceptance (see reviews: Polimeni and Reiss 2003; Brüne 2004). Perhaps the simplest explanation is that schizophrenia is not a traditional medical disease after all. Instead, schizophrenia may represent a vestigial behavioral phenomenon that was once advantageous to ancient hominoids. This resolution has a number of its own problems which will be discussed in later chapters. Leaving aside these other potential obstacles for now, the idea that schizophrenia was once evolutionarily adaptive does seem difficult to fathom. Patients with schizophrenia commonly hear voices that don't exist; they may cling to paranoid beliefs or far-fetched delusions; in its severest form, patients mutter incomprehensibly to themselves. Very few would dispute that, at first sight, the severest symptoms of schizophrenia appear to reflect a gross malfunctioning of normal brain operations, or to put it crassly, a brain gone haywire. How could such a disruptive and seemingly futile behavior *not* be a disease? The laws of evolution, however, seem to challenge our fundamental assumptions about schizophrenia. At the very least, disease models of schizophrenia need to be revisited.

Following this loose thread produced by the schizophrenia paradox led Jeff Reiss and me towards topics foreign to most psychiatrists. As we delved into such fields as anthropology, ethology (the study of animal behavior), group selection, shamanism and the cognitive science behind spirituality, an alternate picture of schizophrenia began to emerge. It was like staring at a Rubin's vase optical illusion and not knowing whether the figure was supposed to represent a vase or face. Flipping from a disease perspective of schizophrenia to a shamanistic explanation necessitated reframing all of what is known about psychiatric disorders — a task much more disorienting than staring at an optical illusion.

Paradigm Shift

I couldn't help but wonder whether this was that "fact" that was preventing us from having a better understanding of schizophrenia. Since 2002, when Jeff Reiss and I wrote our first paper on evolution and schizophrenia, we have both continued to practice psychiatry within the same medical framework as before, treating schizophrenia as a disease. We have also learned to simultaneously appreciate schizophrenia from a completely different vantage point. Although the disease and shamanistic models represent radically different assumptions about schizophrenia, each viewpoint has its own way to explain most of what is known about the phenomenon. Each also has its own loose ends. This is to be expected since few, if any, biological theories are perfectly airtight.

Those readers familiar with Thomas Kuhn's writings will readily appreciate that the shamanistic theory represents a potential scientific "paradigm shift" within the study of schizophrenia (Kuhn 1962; Horgan 1997). Thomas Kuhn wrote a book entitled *The Structure of Scientific Revolutions* (1962) in which he describes the nature of scientific progress. Kuhn rejected the idea that science progresses in an orderly fashion towards absolute truth. Instead, he proposed that scientific advancements were better characterized as a sociological phenomenon. According to Kuhn, scientific observations tend to cluster around a particular perspective. As a scientific field matures, both established and new observations may begin to cluster around another new vantage point. This innovative viewpoint will often solve imperfections (paradoxes or exceptions to the rule) associated with the established theory. Eventually, if the new paradigm seems to make more sense of all the related scientific observations, it will supplant the established theory. Some of the best examples of paradigm shifts are Copernicus's heliocentric model replacing Ptolemy's earth-centered universe, Darwinism replacing Lamarckism, Newtonian physics replacing Aristotelian physics and Einstein's theory of relativity replacing Newton's classical mechanics.

Kuhn, a physicist, came up with the idea of paradigm shifts while preparing for a lecture on the history of Newtonian physics. He was perplexed by that nagging question most of us have when reading the history of science: how could these brilliant historical figures have been so wrong? While preparing for this lecture, Kuhn immersed himself thoroughly into the philosophical world of Aristotle and the ancient Greeks. "I gazed abstractly out of the window of my room. Suddenly the fragments in my head sorted themselves out in a new way, and fell into place together. My jaw dropped" (Gladwell 1996). Aristotle did make sense after all. It was only that Newton — approaching the study of motion from a completely different perspective — made even more sense.

Like the tilting of an old-fashioned two-pan scale, a tipping point is reached when the larger part of a scientific community becomes converted to the new perspective — otherwise known as the paradigm shift. All of the previous observations and facts will be rearranged to fit the new paradigm. Notice there is no underlying supreme logic or absolute standard supporting the new perspective. It just appears more elegant than the older belief system. (The new perspective can also be imperfect with its own paradoxes and exceptions to the rule.)

Although not as earth shattering as those historical examples of paradigm shifts, I will argue that the shamanistic theory of schizophrenia could be a paradigm shift. This is to stress that the idea cannot be easily retrofitted into current disease models of schizophrenia. The reader must be willing to push the reset button and observe schizophrenia from a completely different vantage point. To fully appreciate the new paradigm we must immerse ourselves into scientific areas unfamiliar to most mental health professionals: evolution, group selection, medical history, anthropology, shamanism and the cognitive science behind spirituality. I have not denied any of the conventional facts about schizophrenia — I have only assembled them differently. In fact, almost all of the supplementary evidence for the shamanism theory originates in the contemporary peer-reviewed psychiatric literature.

What the Shamanism Theory is Not

The shamanistic theory of schizophrenia may, to some, suggest a whiff of anti-psychiatric sentiment, but there is no intent to contradict the core medical culture of psychiatry. Winston Churchill's quip about democracy also applies to psychiatry: "It has been said that democracy is the worst form of government except for all the others that have been tried." Even with its checkered history, psychiatry is still among the least ideological of all the institutions that claim to relieve emotional suffering. Even though I am inclined to believe that schizophrenia is, technically, not a disease, this belief hardly affects my own day-to-day psychiatric practice.

Since the 1960s, Thomas Szasz and others have promoted an anti-psychiatric philosophy culminating in the conclusion that schizophrenia, and many other psychiatric disorders, are not medical diseases (Szasz 1974, 1970). Other than coincidentally arriving at similar conclusions, the shamanism theory has absolutely no connection to any of these anti-psychiatric positions. Szasz's denial of mental illness is based on a dogmatic libertarian philosophy taken to such an extreme that it denies the intrinsically social nature of humans.

Some critics suggest that psychiatric theories incorporating shamanism are based on "romantic" notions — implying an underlying desire to idealize spirituality or sanitize mental illness. Many experts in the field do, in fact, possess genuine beliefs in shamanistic practices, and some have even been shamans. I happen to be an atheist, with little personal interest in spiritual matters beyond their possible relationship to the shamanistic theory. As a matter of philosophy, I have never felt particularly compelled to normalize the inherent peculiarities associated with psychiatric conditions. In fact, it is precisely those differences that need to be singled out if one is to properly study psychiatric conditions.

Another criticism is that anthropological or evolution-based theories (including the shamanism theory) tend to be soft or even unscientific. A few years ago I found myself debating this very point with a professor who seemed annoyed that I was spoiling a perfectly good scientific conference with fluffy speculations about mental illness. Driving home later, I replayed that brief contentious debate in my head and concluded that neither of us had precisely articulated the meaning of the word *science*. More troubling was the realization that I had never seen the term satisfactorily defined. How could I defend the shamanism theory as scientific when I didn't know precisely what that meant? Although every person doing scientific work has an intuitive sense of what science is, rarely will two scientists precisely agree on the fundamental definition. To my surprise, it took several weeks of reading and reflection to arrive at a better understanding of what is meant by *science*. Because I have frequently heard this allegation thrown at the shamanism theory, an entire subsection will be dedicated to this issue.

What is Science?

The cultural progress of the last 10,000 years has been astonishing. Compared to humankind's previous accomplishments, this rapid progression of knowledge has been uncharacteristic, perhaps even accidental. It was likely triggered by the advent of farming and animal husbandry, activities that allowed for greater food surpluses and an expanded division of labor. A further acceleration of humankind's ability to acquire knowledge occurred in ancient Greece about 2600 years ago. Science begins here.

Most historians describe Thales of Miletus (c. 624–c. 546 BC) as the first ancient Greek philosopher (Watson 2005). He has also been described as the father of science. Thales lived in the ancient Greek city of Miletus, a thriving commercial city regularly engaged in seafaring trade with several neighboring Mediterranean cities. A number of cultural changes converged in Miletus to create the first truly intellectual environment, and Thales embodied everything that was revolutionary about this enlightened milieu.

Miletus had a regular influx of peoples from diverse cultures including Egypt, Persia and Phoenicia (Laale 2007). The habitual exposure to contrasting civilizations prompted some people to vigorously question their own belief systems and accordingly impressed upon them the arbitrariness of their own customs. Other beliefs, universally shared with other societies, would have appeared more credible. The citizens of these early Mediterranean commercial centers would have been among the first to regularly deal with these two dichotomous forms of information: culture-specific ideologies and universal beliefs. As a consequence, the existence of a knowable universe would have been recognized. The next logical step would have been realizing that thoughtful analysis could conceivably separate real facts about the world from folklore.

Another important addition to the nascent intellectual environment of ancient Greece was the invention of the alphabet. The Greek alphabet arose from the Phoenician alphabet just a few hundred years before Thales was born. Certain scholars stress the superiority of alphabets over earlier pictographic forms of writing, such as Egyptian hieroglyphics, as a primary reason for the ancient Greek cultural explosion (Donald 1991). The uniqueness of each symbol in a pictographic lexicon makes it difficult for most people to remember more than a few thousand words. In contrast, alphabets allow for much larger written vocabularies. Having more words at one's disposal means that the universe can be described in more nuanced ways. Combining alphabets with reams of paper creates the equivalent of an external memory bank, which allows for the analysis of greater amounts of information — similar to how computers are an obvious improvement over pad and pencil.

Still another major development laying the groundwork for scientific thinking may have been the invention of money and the escalating sophistication of entrepreneurial dealings. The first use of stamped coins — made of precious metals — arose in ancient Greece a few decades before Thales was born. Upon learning this detail, I wondered whether this was merely a coincidence. One simple explanation could be that whatever cultural developments were occurring in early ancient Greece, these changes were independently spawning a variety of unrelated inventions. However, I couldn't help think that this daily immersion into an abstract monetary world, with all its many-sided dealings, would have subtly changed how people approached problems. Prior to the use of coins, trading would have typically been tit for tat — a simple two-step sequential relationship. Coinage, on the hand, expands the world of trade into a complex web of multifaceted relationships — good practice for scientific thinking.

It was in this flourishing intellectual milieu that three critical perspectives coalesced, culminating in an attitude we tend to call science: 1) "know thyself," 2) remove supernatural beliefs from philosophical discourse and 3) have a sense that the universe is knowable.

The first perspective underlying scientific thinking is embodied in the aphorism "know thyself" — an expression perhaps first uttered by Thales. To know thyself is to be aware of the strengths and weaknesses of human intelligence. The formal recognition of humankind's intrinsic vulnerability to being fooled would have especially served the earliest entrepreneurs. (It is said that Thales may have been the first to benefit from forming a capitalistic monopoly after purchasing all of the olive-presses right before a particularly good olive harvest.) In a similar manner, in order to understand nature one must creatively solve problems but avoid being deceived by patterns that don't exist; again, understanding humans' capacity to be fooled is valuable. The pursuit of knowledge will be sidetracked into dead ends if one draws bogus conclusions. Resisting beguilement is the purpose of many scientific concepts: the placebo effect, confounding variables, selection bias, statistical validity and Bonferroni correction.

The second important precursor to scientific progress (and arguably an offshoot of "know thyself") is the eradication of folklore or religion from explanations of how the universe functions. In polytheistic ancient Greece, distilling reality from an intertwined pagan folklore would have been the principal task of these original philosophers. It is said that when Thales was asked what he thought about the gods, he replied, "nothing." In his book *The Beginnings of Western Science* David C. Lindberg (2007) argues that the jettisoning of supernatural explanations produced an orderly, predictable and ultimately knowable universe. If the world was a systematic place, then logic was its master.

The third fundamental viewpoint integral to the birth of science was the unwavering belief that a deeper understanding of the world was achievable through thinking. Historians of science, for the most part, agree that this novel outlook, difficult to precisely define, first occurred in these ancient Greeks. The Greeks convinced themselves that through observation and thoughtful analysis, ever more could be known about the universe.

Thales was likely not the first person to question the existence of gods, nor was he the first person to explore his own psychology. That some previously unknown fact can be extracted from nature and built upon must have crossed someone else's mind. However, it is with Thales that these three ideas firmly align themselves for the first time, culminating in a new attitude we now call science.

Although it may be heretical to say, science is more an attitude than a decreed modus operandi. For the human animal, the task of understanding the universe comes down to overcoming cognitive limitations. Many of these limitations involve, for example, our inability to glean patterns without the aid of formal mathematical operations (i.e., statistics). A variety of other operations, such as hypothesis testing, are invoked to minimize the shortcomings of the human brain — always in the ultimate service of elucidating the natural environment. Notice that when the scientific method is applied to man-made objects (including cultural institutions) it is not considered pure science, for example as in engineering, economics, political science, or ethnography.

Another tool of science can be found in the creative expression of every successful scientist. The cognitive mechanics behind innovation, however, are indefinable and tend to be intuitively fostered by each individual. Science is indubitably facilitated by imagination, but because creativity is intangible, no set of formal rules exists to lead scientists to an ideal creative state of mind. It is not, however, farfetched to suggest that some day scientists will be able to methodically place themselves into the most inventive mindset possible — and that such mental approaches could be formalized inside the scientific method.

Too many scientists erroneously believe that the individual tools of science, such as hypothesis testing or falsification, reflect the essence of science. For example, the technique of hypothesis testing is often used simply to compensate for humankind's less than perfect memory. The apocryphal story of Galileo throwing objects off the Leaning Tower of Pisa will serve to demonstrate this point. If all people were able to recall with infallible accuracy that lighter objects fell at the same approximate speed as heavier objects, an experiment of this kind would have never been necessary. Instead, the results of Galileo's experiment would have been axiomatic rather than a dubious opinion requiring confirmation. In fact, if hypothesis testing were so crucial, Isaac Newton should have dropped several hundred apples onto his head so as to statistically confirm the constancy of gravity.

Similarly, the acclaimed science philosopher Karl Popper mistakenly placed the idea of scientific falsification as the centerpiece of science (1959). Instead, falsification is just another tool to test the integrity of a theory — nothing more. As critics of Popper have pointed out, falsification has no greater intrinsic validity than verification. In other words, scientists can just as easily make a mistake believing they have falsified an acceptable theory. Falsification is simply not the be all and end all of science. The dissident philosopher of science, Paul Feyerabend, may have been on to something when he said that in science "anything goes" (Godfrey-Smith 2003).

So, by my account, there is nothing intrinsically unscientific about the shamanistic theory of schizophrenia. I hope to convince each reader that the central arguments of the theory are reasonable and unencumbered by gross distortions or blatant dogma — in short, legitimate science (and for followers of Karl Popper, the shamanistic theory can generate testable hypotheses — see chapter 10 for a pilot study). Although the shamanistic theory may not fit together perfectly (neither do established disease models), I believe the theory offers, at the very least, the potential to further our understanding of schizophrenia. Almost all of my patients with schizophrenia have wished that this thing, whatever it is, had never happened to them. With millions of people struggling with this often-terrible problem, no stone should be left unturned.

– 2 –
The Sculpting of Schizophrenia

This chapter will provide readers who may be unfamiliar with psychiatry a better sense of the perspectives that dominate the field — especially those views that impinge on the shamanistic theory of schizophrenia. The concept of schizophrenia is neither an entirely man-made social construct nor a pure reflection of a natural phenomenon. Schizophrenia is like a marble statue: there is a natural core but the shape has been chiseled out by the scientific community. Our biology provides the stone but Western culture shapes it. This is not to diminish the stature of the Western formulation of insanity; it has taken many clever minds to discover everything that is known about schizophrenia. However, the precise modern conceptualization of schizophrenia has been fashioned by both cultural idiosyncrasies and little historical turns. The first part of this chapter reviews the history of psychiatry while the second section, The Faces of Psychosis, provides a basic description of psychosis and schizophrenia.

A Brief History of Psychiatry

People are undeniably shaped by history. Certain political details of European history, for example, can explain why Northern Europeans practice Protestant Christianity and Southern Europeans turn to Catholicism. Similarly, the practice of psychiatry is undeniably influenced by the meandering of its historical path. A little background is therefore necessary to understand how psychiatrists understand the mind and how the modern conceptualization of schizophrenia came to be.

The first rudimentary psychological therapy was perhaps an extension of Buddhism, which originated in India around the sixth century BC. Its teachings, lent weight by spiritual ideas such as karma, provide a collection of positive outlooks that tend to mitigate suffering, bolster happiness and foster social harmony. It can still be useful to remind oneself that inflated desire can lead to discontent, one of the fundamental tenets of Buddhism.

While the foundations of Buddhism (and Confucian philosophy in China) were being laid down, ancient Greek philosophers were attempting their first forays into the psychology of the mind. Plato's separation of the rational versus irrational soul can be regarded as the first formal division between emotions and purely logical thoughts (Stone 1997, 8). Although prehistoric peoples must have had a sense that emotions such as anger, sadness or love can cloud judgment, it is not until humankind consciously disentangled emotions from logic that the study of each respective facet of mind could begin to occur.

Hippocrates (460–370 BC), also known as the father of medicine, made the seminal contribution of unambiguously removing magical or spiritual beliefs from explanations of illness. Using the ancient Greek conception of physical matter, Hippocrates speculated that corporal imbalances such as excessive heat, cold, dryness or wetness could cause emotional ailments like madness or delirium (Millon 2004, 15). Epictetus, another ancient Greek philosopher, laid down the basic principle of cognitive therapy almost 2000 years ago when he said, "People are disturbed not by things but by the view which they take of them" (Hunt 1993, 675).

The Renaissance philosophers were the next group to substantially advance psychological theory. Beginning with Niccolo Machiavelli's (1469–1527) unyielding cynicism about political discourse, the Renaissance intellectuals began to delve further into the nuances of thought. Combining introspection and observation, philosophers became the psychologists of their day. Scholars such as Francis Bacon, Benedict Spinoza, Thomas Hobbes and John Locke made astute observations about human nature and philosophized about the political, psychological and sociological questions of their time. Further speculation on how the mind works continued with the Enlightenment philosophers of the eighteenth century, such as David Hume and Emmanuel Kant (Hakim 1967).

Some philosophies of the mind were quite sophisticated and elaborate. For example, in the late 1700s the little-known Scottish philosopher James Beattie propounded a theory on the psychological elements of humor (Beattie 1776). He was well aware that humor appeared to be caused by two "incongruous parts" forming an "assemblage" in the mind. Not until the 1970s were cognitive scientists able to make any appreciable enhancements on Beattie's formulation.

By the late nineteenth century the technique of combining social observation and introspection in order to understand the mind was beginning to hit a dead end, or to be more accurate, producing diminishing returns. The theories of philosophers such as Emmanuel Kant and Friedrich Nietzsche were becoming increasingly esoteric and inscrutable, and as such this sort of inquiry began to fall out of favor. Whether by necessity or coincidence, the late nineteenth century saw the study of the mind divide into three paths: experimental psychology; psychoanalysis (which ran into its own dead end after a few decades); and the rise of insane asylums and the medicalization of psychosis.

The experimental psychologists can trace their roots back to Darwin's two most important books, *On the Origin of Species* (1859) and *The Descent of Man* (1871). With a one-two punch, Darwin's works inescapably linked animals to humankind, dispelling the idea that the human mind was distinct and special. Two centuries before Darwin, Rene Descartes' influential philosophy of dualism had asserted the uniqueness of humankind. Descartes proclaimed that the human mind consisted of two components: the physical matter — the brain — and a non-physical aspect — the soul. While human beings consisted of physical matter plus a soul, animals were soulless; merely complicated organic robots. In contrast, the philosophy of materialism or physicalism (which has its roots in the ancient Greek atomistic philosophy of Leucippus and Democritus) asserts that only matter exists, so spiritual or other intangible feelings are only in our imagination. In other words, humans too are merely complicated organic robots. Darwin's evolutionary theory, in conjunction with the philosophy of physicalism, helped discard the "ghost in the machine."

Removing the soul from the equation had the effect of simplifying the mind — if only conceptually — which freed scientists to try and understand it through experimentation. The experimental psychologists of the late nineteenth and early twentieth century began to abandon introspection and focus their efforts on measurable parameters: stimuli and behaviors. The mind was treated as a black box with quantifiable inputs — sensory stimuli such as light, sound, or touch — and outputs — behaviors such as reaction time or bar-pressing frequency.

In 1879, Wilhelm Wundt built the first laboratory dedicated to psychological research, at the University of Leipzig (Hunt 1993). A few years later in Russia, the physiologist Ivan Pavlov made a serendipitous discovery while studying gastric function in dogs (Pavlov 1955). Pavlov noticed that the dogs were salivating *before* their food arrived. The dogs were being tipped off by the approaching footsteps of their handlers, and had unconsciously learned that mealtime was imminent.

Pavlov immediately understood that there was something automatic and fundamental about this type of learning. Through meticulous experimentation, his lab established the basic tenets of classical conditioning (or Pavlovian responses), one of the major building blocks of how animals, as well as people, learn.

Psychologists began to explore various forms of learning through controlled animal experimentation. Edward Thorndike used puzzle boxes to observe how cats escape from various forms of confinement. His observations led to some of the first suppositions about how ideas become associated with each other. In the 1930s B. F. Skinner began to study how animals modify their behaviors depending on how the environment rewards or punishes their actions. Being mechanically minded, Skinner was able to build a sophisticated device that could control the delivery of a reward (food pellets) while accurately measuring consequent behaviors (usually pressing a bar). Although the device was originally called an "operant conditioning chamber," it later became known as a "Skinner box."

The type of learning happening inside the Skinner box was called operant conditioning. Whether he studied rats or pigeons, there were universal patterns to how animals learned the consequences of their actions. For example, altering the timing and reliability of rewards predictably affected how quickly animals learned. More importantly, these basic principles of operant conditioning also applied to human beings.

In the 1930s, Nikolaas Tinbergen, Karl von Frisch and Konrad Lorenz began to methodically observe animal behaviors in the wild. They shared the 1973 Nobel Prize in Physiology or Medicine for discovering a number of principles underlying the social behaviors of animals, and for establishing the field of ethology, the formal study of animal behaviors in the wild. Von Frisch was particularly well known for his studies related to insect communication and for interpreting the meaning of the honeybee waggle dance (von Frisch 1953). Lorenz, studying geese, became famous for his keen description of imprinting, a form of instinctive time-sensitive learning that typically occurs in newly hatched chicks. Both Lorenz and Tinbergen laid the groundwork for such concepts as *supranormal stimuli* (artificial stimuli that prompt exaggerated responses, like junk food or pornography) and *fixed action patterns* (instinctive behavioral sequences, e.g., mating dances, beaver dam construction).

John Bowlby, a British psychoanalyst, drew from the work of the ethologists and began applying it to his patients. Bowlby developed *attachment theory* to explain how children become connected to their parents and how this colors their social relationships throughout life (Bowlby 1988).

While the experimental psychologists were entering the mind using the front door, psychoanalysts were trying to slip in through the back. Sigmund Freud, the creator of psychoanalysis, did most of his original work in the 1890s and early twentieth century (Alexander and Selesnick 1966; Brenner 1955; Brill 1938; Gay 1998; Watson 2006).

Freud's unique insight was that the unconscious mind dictates human behaviors more than anyone previously imagined. Freud was a psychiatrist who saw normal human existence as a caldron of unconscious desires battling it out, with the winner steering the next conscious thought or palpable behavior. Although Freud may have overestimated the fiery intensity of this conflict or the extent of its libidinous content, no one before him had ever attempted to systematize the activities of the unconscious mind in such a thorough and methodical manner. Many popular psychodynamic terms originate with Freud: Oedipal complex, pleasure principle, repetition-compulsion, penis envy, free association, anal personality, id, ego and super-ego.

In addition to being an unapologetic atheist, Freud ardently believed in psychological determinism — the idea that the stirrings of the mind aren't random but that every thought is logically connected to antecedent thoughts. Although billiard balls may look like they've been splattered haphazardly after the initial break, in fact each ball's placement is precisely determined by the speed, angle and location of the impact of the cue ball. Freud believed that, similarly, if you could learn enough about the mind you would be able to predict its next move — accordingly, free will does not exist. It was this philosophy that gave Freud the confidence to delve into the elusive world of the unconscious mind.

Freud created a number of techniques to plumb the unconscious mind, including having patients free-associate, examining slips of the tongue, and analyzing dreams. Dream interpretation was Freud's preferred method; he dubbed it the "royal road to the unconscious."

Freud also laid down the basic principles of psychotherapy, or as one of his patients called it, the "talking cure." The idea of an expert listening to, analyzing, and interpreting the problems of an emotionally distressed person was first formalized with Freud. The neurotic sufferers of the twentieth century, mostly members of the American and European upper-middle classes, were powerfully attracted to this form of therapy. (It is possible that repeatedly talking about oneself may not have been helpful to the emotionally distressed people of other epochs.)

Most of Freud's finer points in support of theories were highly speculative and probably off the mark. However, a few of his therapeutic observations have survived and remain useful to this day. For example, Freud postulated that the mind has a variety of tricks, known as defense mechanisms, to impede or repress anxiety-provoking thoughts from entering conscious awareness. Sometimes these defenses can be helpful while at other times harmful, resulting in pathological behaviors. Denial is just one of about twenty defense mechanisms surmised to exist. Freud's list of defense mechanisms remains in the vocabulary of almost every contemporary psychotherapist.

Transference is another valuable therapeutic concept originally conceived by Freud and developed more fully by subsequent psychoanalysts. Transference describes the underlying feelings that color the relationship between patient and therapist. Freud noticed that patients sometimes made assumptions about him that were patently false. These misconceptions seemed to have more to do with childhood conflicts than anything else. For example, occasionally a patient will become unjustifiably annoyed with their psychotherapist — particularly a patient who was neglected or abused by their parents. Psychoanalytic theory posits that anger towards a parent will be more painful (due to feelings of guilt) than anger towards the psychotherapist. Supposedly, residual hostile feelings lie dormant in the unconscious mind until unintentionally stirred by a psychotherapist, whose position of authority is reminiscent of that of a parent. The pressure of accumulated frustration is released and directed at the therapist, presumably due to their symbolic resemblance to the parent; hence the transference. An open discussion of buried resentment towards parental figures often dissipates the anger that was inappropriately directed at the therapist.

The concept of transference hints at the richness of the unconscious mind, but it can also serve to expose the potential fallacies of psychoanalysis. Having the patient's fury dissolve soon after supplying a sensible interpretation would seem, on the surface, to be evidence that the therapist was on the right track. However, less obvious but equally viable explanations exist for this sort of patient-therapist encounter. What if the patient's rage is actually constitutional or hereditary, and not truly fueled by conflict? What if every person, to varying degrees, possesses some concealed antagonism towards their parents? Perhaps the therapist's detailed interpretation of the patient's misplaced anger actually serves as a metaphor about hostility in general; the patient takes this new knowledge about the psychological mechanics of anger and unwittingly applies it to the therapy session. In other words, the supposed primeval parental conflict could have been relatively meaningless.

By the 1950s, psychoanalytic theorists became increasingly imaginative and generated dozens of wide-ranging psychodynamic interpretations for any given clinical scenario. This produced mounting skepticism in all but a minority of psychiatrists and caused the gradual rejection of esoteric psychoanalytic theory.

We have thus far examined two of the three disciplines that shaped contemporary psychiatry: experimental psychology and psychoanalysis. The third group to heavily influence modern psychiatry were the physicians who oversaw the insane asylums of the late nineteenth century. These pioneers made formidable contributions towards understanding the mind, especially at the juncture where normal brain functions interface with psychosis.

Although a smattering of small insane asylums existed throughout the Middle East and Europe during the Middle Ages, it wasn't until the nineteenth century that the phenomenon of the mental hospital really took root in western Europe. The eighteenth century Enlightenment brought the social changes that sparked the French Revolution, as well as greater investment in public institutions throughout Europe. As Europe grew wealthier, physicians began taking over the custodial care of the incarcerated mentally ill. By the late 1700s, Vincenzo Chiarugi in Italy and Philippe Pinel in France advocated for more humane treatment of the mentally ill, including eliminating chains from asylums (Shorter 1997). By the late 1800s, university research programs were beginning to be closely affiliated with mental hospitals.

The sickest psychotic patients began to be regularly supervised by physicians, who plied their medical training towards observing and trying to understand insanity. The 1800s spawned almost as many theories about insanity as there were psychiatrists. To nineteenth century psychiatrists, psychotic symptoms were so haphazard that no obvious pattern emerged. (Contemporary psychiatry is probably not faring much better.) Delusions and hallucinations seemed to come in an inexhaustible number of configurations. To make matters worse, each patient's social milieu colored psychotic symptoms, but not always in predictable ways. Because most psychotic patients showed no physical pathology such as brain lesions or abnormal blood tests, the problem of categorizing mental illness looked bleak.

Emil Kraepelin was among those first academic psychiatrists who devoted their careers to understanding psychosis. Born in Germany in 1856 (the same year as Freud, although the two men never met), Kraepelin initially studied with professor Wilhelm Wundt, one of the pioneers of experimental psychology. Kraepelin became a meticulous observer of his psychotic patients, who numbered in the thousands. He eventually divided psychosis into two types: psychosis in the absence of emotion, and mood-ridden psychosis. He labeled psychosis in the absence of emotion *dementia praecox* (premature dementia); it's now known as schizophrenia. Mood-ridden psychosis was called manic depression and is now termed bipolar disorder.[1]

The way that Kraepelin supported his diagnostic classification was unique and ingenious (Palha and Esteves 1997; Greene 2007; Berrios and Hauser 1988). Kraepelin's leap of faith was that insanity was a biological disease — even though his two capable neurohistologist colleagues, Franz Nissl (inventor of Nissl microscopy staining) and Alois Alzheimer (discoverer of Alzheimer's disease) had not found any associated neuropathological findings. In the absence of pathological findings, a proxy measure was needed. Kraepelin thus turned to prognosis — assessing a patient's symptoms and forecasting their probable course.[2] The detailed records from Kraepelin's clinic indicated that mood-absent psychosis tended to follow a deteriorating course of illness, while mood-ridden psychosis usually resulted in a more stable course. Kraepelin sensibly

---

[1] In the 1850s, manic depression was described by two Frenchmen and termed "circular insanity"; however, it represented one of several subtypes of psychosis (Baethge et al. 2003).
[2] Karl Kahlbaum too had previously entertained the notion of using prognosis to differentiate mental illness (Berrios and Kahlbaum 2007; Brüne 2000).

presumed that the divergence in prognosis was the result of two different pathological mechanisms. It was this extra piece of evidence that sold his already appealing, simple-to-use, two-category scheme (Angst 2002).

The term *schizophrenia* was coined by another influential German-speaking psychiatrist, Eugen Bleuler (who was Swiss), in the early 1900s.[3] Bleuler had a reputation for spending a lot of time with his patients and as a result began noticing much more than just bizarre behaviors. He believed that the essence of schizophrenia resided in four subtle yet enduring symptoms: inappropriate affect, extreme ambivalence, loose associations and autism (also known as Bleuler's four As). Psychosis, because of its unpredictability, was relegated to the status of auxiliary phenomenon. It made sense to Bleuler that such permanent symptoms as loose verbal associations or autism were likely to be more integral to schizophrenia than the often-transient psychotic symptoms. Relegating the most striking symptom of an illness to the sidelines could be construed as arbitrary but Bleuler's reasoning was, at least, defensible.

Between Kraepelin and Bleuler, psychotic symptoms were losing their primacy. The spotlight, however, shifted back to psychosis after the introduction of the first truly effective schizophrenia medications in the 1950s. Antipsychotic medications can sometimes completely eliminate psychosis; however, they are only marginally effective against Bleuler's enduring symptoms.[4]

We have very briefly reviewed the three disciplines that represent the foundations of modern psychiatry: experimental psychology, psychoanalysis and the medical treatment of the insane. The next great seismic event to shape contemporary psychiatry was the development of effective pharmacological treatments.

Prior to World War II there were no satisfactory treatments for the insane. A supportive environment or compassionate listener could have a calming influence, but little if any, curative effects upon the psychosis itself. A number of medicines, like barbiturates, were tried but at best they could only sedate. In the 1930s lobotomies, electro-convulsive therapy and insulin coma treatments were used for the most severely mentally ill. Although each of these therapies was effective to some extent, they also had serious drawbacks. Contemporary writers have ridiculed these treatments, but the modern reader must keep in mind that many asylum patients were suffering with unbearable psychotic symptoms and caretakers were desperate to try anything at their disposal.

---

[3]Schizophrenia literally means "split mind," which described Bleuler's view that thoughts and emotions were separated.

[4]It must be acknowledged that the fact that antipsychotics can often eradicate psychosis but do not resolve the four As is consistent with Bleuler's formulation.

The first truly effective psychiatric medications were developed in the 1950s and thereafter, therapeutic interventions in psychiatry began to shift their focus; psychiatrists started routinely using a combination of talking therapies and drugs. The first use of an antipsychotic medication was the result of a serendipitous discovery by two French psychiatrists, Jean Delay and Pierre Deniker. They discovered that chlorpromazine, a compound that had initially been intended as a pre-anesthesia agent, could cure psychotic patients almost like magic.[5] The 1950s also saw the development of the first truly effective antidepressants, as well as the gradual introduction of lithium as a mood stabilizer for bipolar disorder. Some years later, a few anticonvulsants were also found to be useful in blunting mania.

Over the last few decades, dozens of antipsychotics and antidepressants have hit the market. The newer atypical antipsychotics (i.e., olanzapine, risperidone, quetiapine) came into the market in the 1990s and although not any more effective than the older antipsychotics, were often subjectively better tolerated. Clozapine may perhaps be the best treatment for schizophrenia and schizoaffective disorder but it is also the most potentially toxic schizophrenia drug, thus limiting its use. In 1986, Prozac (generic name fluoxetine) changed the focus of depression treatment by broadening the clinical population willing to undergo such treatments. Compared to the older antidepressants, Prozac had substantially fewer side effects, as well as a wider range of medical uses. (It could treat milder forms of anxiety and depression as well as obsessive-compulsive symptoms.) As a consequence, greater segments of the population were electing to try Prozac and other newer antidepressants like Zoloft, Effexor, Celexa and Paxil.

Although pharmacotherapy has become a major constituent of psychiatric treatment, talking therapies still form an integral part of most clinical practices, especially for those emotional problems that used to be known as the "neurotic disorders": chronic low grade depression and borderline personality traits. Although a countless number of specific therapies have been authored, almost all psychotherapies fall into three general categories: supportive psychotherapy, insight-oriented psychotherapy and cognitive-behavioral therapy (CBT).

Supportive psychotherapy is a treatment that attempts to be reassuring and empathetic towards patients. Technically speaking, supportive psychotherapy seeks to strengthen a patient's healthy defense mechanisms.

Insight-oriented psychotherapy has replaced traditional psychoanalytic therapy for most therapists. The basic premise is to help patients

---

[5]Incredibly, the drug company that owned the patent was at first disinterested in its psychiatric use because they doubted whether there would be a market for psychiatric drugs. (Healy 2002, 92). Antipsychotics sales have since become a multi-billion-dollar industry.

explore their emotional problems through a better understanding of general psychological principles. Insight-oriented therapies tend to be less esoteric than psychoanalytical approaches.

Cognitive behavioral therapy, developed by Dr. Aaron Beck, encourages rational thinking to negate the cognitive distortions frequently seen in depressed or anxious patients. For example, "I am a failure" is a feeling commonly observed in depression. CBT asks the patient to systematically examine and challenge this sort of arbitrary pessimistic thinking.

Most psychiatrists are familiar with the basic teachings associated with each type of psychotherapy and informally apply these ideas to their clinical cases. Psychotherapy alone is appropriate for mild depression, relationship problems, grief and personality disorders, to name a few. For moderate depression, a combination of psychotherapy and medications has been found to be slightly superior to each treatment by itself. For other psychiatric problems, like schizophrenia and bipolar disorder, medications have a much greater relative role than psychotherapy.

The last major transformation in psychiatry occurred in 1980 with the development of DSM-III, the third edition of the Diagnostic and Statistical Manual of Mental Disorders. Compiled by the American Psychiatric Association, the DSM lists all the possible psychiatric diagnoses according to a consensus of experts from the international psychiatry community. The third edition of the DSM was different from preceding versions: Robert Spitzer, the DSM-III's main architect, insisted on operational criteria for each diagnosis. In other words, every psychiatric diagnosis would include a checklist of associated signs and symptoms. A psychiatric diagnosis could only be made if the patient fulfilled enough of the listed criteria. In this way, psychiatric diagnoses would eventually be more precise and less subjective.

However, the DSM, now about to issue its fifth edition, has a number of limitations (First 2010). Foremost, it is a diagnostic classificatory system with no theoretical underpinnings. In other words, physiological mechanisms are given little consideration. The current classification of psychiatric diagnoses is based on not much more than hunches, making it arbitrary and fallible. In addition, the question of whether a specific criterion associated with a diagnosis is present in the patient can be subjective, leading to differing diagnoses. Having said this, the DSM has been enormously useful to psychiatry for two reasons: because it creates a common language for psychiatrists, and because refining operational criteria can conceivably lead to useful psychiatric diagnoses.

The DSM criteria are like genes. Diagnostic criteria are under continuous study and those that inadequately correlate with their respective diagnoses will not survive to make the next version of DSM. Meanwhile, newly proposed operational criteria (mutations) may be added to upcoming editions. Darrel A. Regier, involved in the planning of the next DSM manual, says that it is their intention that DSM-V should be a "living document" (Tamminga et al. 2010). Perhaps we have come full circle — Darwin's great discovery kicked off the psychological sciences and now, 150 years later, the DSM planners are unwittingly applying evolutionary principles to developing psychiatry's principal doctrine.

The Faces of Psychosis

Like the human face, each person's expression of psychosis is unique but also shares a gestalt with others. The possible archetypes of psychosis, however, have not yet been brought into focus. Although we have discerned some general patterns associated with psychotic thinking, these ideas are not at all tethered to any elementary aspect of normal psychological, sociological or neurological function. Put another way, conventional psychiatric theories provide no sense of purpose to psychosis. Even if insanity turns out to be a disease, there is still no widely accepted theory describing which aspect of normal psychological function has been derailed.

The words insanity and psychosis are almost interchangeable; psychosis is the more commonly used medical term. Psychosis can be expressed in one of three ways — delusions, hallucinations or disorganized thinking — but the basic premise of psychosis is being out of touch with reality. But out of touch in a certain way; one may be deluded but not psychotic. For example, there are people who are deluded because of dogmatic philosophies, lack of intelligence or immaturity, but they are not necessarily insane. Likewise, hallucinations can occur in grieving individuals or people who hear their names called out in a crowded mall. Thus, insanity cannot be so easily reduced to a simple definition and instead seems to represent a richer multifarious process.

There are a dozen or so purported causes of psychotic thinking. Each hypothetical root cause produces its own distinctive flavor of psychosis. And like flavors, the different forms of psychosis are sometimes difficult to tease apart. There is not enough specificity between the various forms of psychosis to reliably distinguish them from each other. Schizophrenia, for example, can be associated with almost any kind of psychotic symptom, but is most often characterized by bizarre delusions, auditory hallucinations and paranoid thoughts. In bipolar disorder, delusions tend to be less unusual, often to the point of being plausible. Bipolar delusions tend to follow the respective moods: nihilistic thoughts during depression and grandiose ideas during mania. Delusional disorder is typically represented by a single non-bizarre delusion without hallucinations or disordered thinking. Significant childhood abuse or neglect can occasionally lead to hallucinations in adulthood, although this type of psychotic expression tends to be short-lived and impose less upon the patient's perception of reality.

Medical conditions such as brain tumors or temporal lobe epilepsy (TLE) sometimes produce psychotic experiences. The temporal lobes are heavily involved in auditory signal processing, which may explain the preponderance of auditory hallucinations in TLE psychosis. Metabolic insults like hepatic diseases or infection can induce delirium, an impaired mental state characterized by obtundity and confusion as well as hallucinations or delusions. Neurodegenerative disorders such as Alzheimer's dementia can also be accompanied by misattributions, hallucinations and delusions. Confusion secondary to delirium or dementia will often emerge in environments of reduced stimulation, such as late evening. This late-day confusion is so common that it has its own term: sundowning.

Drug-induced psychosis is self-limiting and usually does not present to physicians. Dozens of substances like LSD, amphetamines, marijuana and cocaine can readily cause psychosis. Each psychoactive drug generates its own brand of psychosis. For example, psychedelic drugs, such as LSD, tend to bring about visual hallucinations. Marijuana frequently causes paranoia. Cocaine energizes its users, sometimes provoking disordered thinking as well as grandiose or paranoid psychotic themes. One of the most bizarre psychotic episodes I ever witnessed involved a cocaine-intoxicated patient having a very animated conversation with an old-fashioned upright blood pressure apparatus. In contrast, patients with schizophrenia usually just talk to themselves.

Of all the manifestations of psychosis, schizophrenia is unquestionably the diagnosis most associated with psychotic symptoms. The symptoms of schizophrenia are sometimes divided into two sets: positive and negative. Psychotic symptoms such as delusions, hallucinations and disjointed thoughts are considered positive, because they are not normally present in healthy individuals; negative symptoms are related to traits that are normally present in healthy people but absent in people with schizophrenia. Negative symptoms include poor motivation, lack of affect, social withdrawal, poverty of speech, anhedonia and mild cognitive impairment (recall Bleuler's subtle yet enduring symptoms). There may be something to this arbitrary division because anti-dopaminergic chemicals (antipsychotic medications) seem to preferentially modify positive symptoms compared to negative symptoms.

Auditory hallucinations are one of the most common symptoms of schizophrenia. Voices can be either male or female but precise identities are often obscure. It is rare that a voice represents someone the patient knows very well. In many instances, hallucinations convey divine or diabolical messages. Auditory hallucinations can reflect warnings or criticism of the patient. Sometimes they command patients to do something, even kill themselves. Most patients find the voices disturbing or at least a nuisance; however, there is the occasional patient who enjoys hearing voices. Visual hallucinations tend to be less prevalent in Western societies.

Delusions are another very common symptom associated with schizophrenia. Psychotic delusions rarely augur something pleasant. Instead, they usually warn of impending trouble and accordingly, paranoid delusions are very common. Having paranoid delusions does not necessarily mean one is suspicious in character. Rather, paranoid delusions usually feature a limited set of completely conjured up perils — for example, demons, street gangs, police or aliens.[6]

Many delusions involve the feeling that actions or thoughts are being controlled by other entities: anonymous people, media personalities, interplanetary life forms or spiritual beings. Telepathic communication from spiritual or other unfathomable entities is a common experience. Inconsequential occurrences can take on monumental meanings; one patient, upon hearing any high-pitched sound, believed foreign thoughts were being put inside his head. Another patient was convinced she was going to die every time she heard an ambulance in the distance. Sometimes, delusions can be downright preposterous, such as a patient who sincerely asked me if I knew God's fax number or whether I had any information about "the schizophrenic planet."

Disordered thinking is usually associated with the most severe cases of schizophrenia. Patients may mumble to themselves, giggle for no apparent reason or blurt out angrily. Disorganized psychotic thoughts resemble a collage of unrelated ideas as opposed to a linear narrative.

A number of identifiable behaviors have been associated with schizophrenia. Patients with schizophrenia often lack facial expressivity; animated expressions are replaced by a flat, robotic affect with very little blinking. Basic grooming can be neglected; some patients won't comb their hair, brush their teeth or even change their clothes. On rare occasions, people with schizophrenia can become catatonic — psychologically unresponsive to their environment — for hours at a time. They may even assume strange and seemingly uncomfortable postures for hours. One of the most tragic behavioral outcomes of schizophrenia is suicide. The modern suicide rate for patients with schizophrenia is about 5%, which

---

[6] A colleague once told me about a curious interview he had with a young man with schizophrenia whose main problem was his numerous and outlandish paranoid delusions. The psychiatrist, meeting the patient in his hospital room, noticed that a hospital security camera had just swiveled around, seemingly peering into the room. My colleague naturally asked his patient if had concerns about what was happening outside his window, but the young man matter-of-factly replied, "No doc, that's just a security camera."

is significantly higher than the norm (Palmer et al. 2005; Hor and Taylor 2010).

Most patients with schizophrenia reside in group homes, low-income housing or with their family. They typically have few friends or sometimes none at all. Despite this relative social isolation, people with schizophrenia rarely complain of loneliness. Unemployment rates exceed 80% in most Western jurisdictions. Very few people with severe mental illness have stable marriages, raise children or work independently. This description applies to Western societies; schizophrenia tends not to be as debilitating in less technologically developed societies.

– 3 –
The Schizophrenia Paradox

The field of evolutionary psychiatry came into existence about fifty years ago with the publication of a seminal paper by Julian Huxley, Ernst Mayr, Humphry Osmond and Abram Hoffer (Huxley et al. 1964). This brief hypothetical piece was the first attempt to formally apply evolutionary principles to psychiatric disorders. According to the article, Julian Huxley (grandson of T. H. Huxley, the biologist known as "Darwin's bulldog") and Ernst Mayr, one of the early pioneers of evolutionary theory, had each independently arrived at similar conclusions about the relevance of evolutionary theory to schizophrenia. They identified two characteristics of schizophrenia that seemed incompatible according to evolutionary theory: schizophrenia's high prevalence rate of about 1 percent, and the low fecundity of schizophrenia patients. They questioned how schizophrenia, a condition with an appreciable heritable component, persists if those afflicted consistently have fewer progeny?

This troublesome anomaly is now known as the *schizophrenia paradox*. The resolution proposed by Huxley and his colleagues was that schizophrenia genes might confer resistance to wound shock and infections, and protection against abnormally high levels of insulin and other hormones. This speculative solution has never been substantiated and is almost certainly mistaken. A dozen or so researchers have since proposed alternative evolution-based solutions to the schizophrenia paradox: one of them is the shamanistic theory (see reviews: Polimeni and Reiss 2003; Brüne 2004).

The schizophrenia paradox implies that there is a beneficial aspect to schizophrenia — otherwise, how would natural selection support it? It implies further that schizophrenia is not a medical disease in the conventional sense. There are several suppositions embedded in the schizophrenia paradox which require closer examination:

i. Schizophrenia is heritable: it is primarily a genetic condition.
ii. The prevalence of schizophrenia exceeds its mutation rate: there are more cases of schizophrenia than you would expect if it were caused only by random genetic mutation.

iii. Schizophrenia is associated with reduced fecundity: people with schizophrenia have fewer children than the rest of the population.
iv. The incidence of schizophrenia has been relatively stable through history and through the world.
v. Schizophrenia is an old phenomenon — its age is measured on evolutionary timescales.

The first section of this chapter will review the heritability of schizophrenia. The next section will examine the basic epidemiology of schizophrenia, including its prevalence, the associated reduction in fecundity, and its geographical and temporal stability. The third section will consider how far back in time schizophrenia can be traced.

The Heritability of Schizophrenia

A variety of findings support an appreciable genetic role in schizophrenia. It has long been observed that schizophrenia tends to run in families. As far back as the early 1900s, Kraepelin noticed that the majority of his patients with schizophrenia had family histories of major psychiatric illness (Shorter 1997, 240). Family studies, however, do not prove genetic affiliation, because environmental influences affecting entire families could contribute to the psychiatric disorder. To tease out potential contributions from the environment, two types of studies have been employed: twin and adoption studies.

Identical, or monozygotic, twins are generally considered to have matching genetic compositions while fraternal, or dizygotic, twins share on average 50 percent of their genes. The genetic differences between monozygotic and dizygotic twins can be used to approximate the heritability of a medical condition. In the 1940s Franz Kallmann surveyed 691 sets of twins of which at least one twin was diagnosed with schizophrenia. He found an 85 percent concordance rate for schizophrenia among monozygotic twins and only 15 percent concordance for dizygotic twins (Kallmann 1946). Modern studies, using stricter diagnostic criteria, have yielded similar results: according to Cardno and Murray (2003), the pooled concordance rates for the four most recent schizophrenia twin studies using DSM-III-R criteria were 50 percent for monozygotic twins and 4 percent for dizygotic twins.

Adoption studies are another way to estimate the genetic factors underlying any given phenotype. Seymour Kety used records from the Danish national adoption registry, known for its comprehensiveness, and found that first-degree biological relatives of adopted children with schizophrenia had substantially higher rates of schizophrenia than control adoptees (Kety et al. 1994). This data was later reanalyzed using DSM-III criteria, yielding comparable results: looking at the adoptees with schizophrenia, 8 percent of their first-degree biological relatives had schizophrenia compared to 1 percent for control adoptees (Kendler et al. 1994). Recent Finnish adoption studies reveal similar ratios (Tienari and Wynne 1994; Tienari et al. 1994; Tienari et al. 2000).

Heritability is a measure of the phenotypic variance attributable to genes as opposed to the environment. Estimating heritability can be extremely complicated because calculations are based on a number of sometimes tenuous assumptions. Moreover, analyses of heritability do not take into account possible gene-environment interactions. Even so, recent studies suggest that the heritability of schizophrenia could be as high as 85 percent (Cardno et al. 1999).

Illuminating the precise genes associated with schizophrenia has been more difficult than initially anticipated. Over the years there have been a few claims of the discovery of "the schizophrenia gene," but nothing has ever panned out. There are, however, several polymorphisms — with connections to neurophysiological functions — that are beginning to show putative associations with schizophrenia (Harrison and Weinberger 2005). Polymorphism is the physical or behavioral difference observed between organisms of the same species, presumably due to differences in underlying genes. For example, eye color has at least three polymorphisms: blue, brown, and hazel. Some researchers have estimated that ten to thirty genes could, in one way or another, contribute to a person's vulnerability to schizophrenia. Technological advances in genetic research are happening at a rapid pace and as a result, medical researchers may finally be getting closer to cracking the genetics of schizophrenia. (Chapter 4 will review the genetics of schizophrenia in greater detail.)

The Epidemiology of Schizophrenia

The Prevalence of Schizophrenia

One of the most critical suppositions pertaining to the schizophrenia paradox is that the prevalence of schizophrenia exceeds typical mutation rates. Prevalence is the number of cases occurring in a population at any given period in time, usually expressed as a percentage. The prevalence of schizophrenia has been extensively studied throughout many cultures and is usually around 1 percent (McDonald and Murphy 2003) — a figure unmistakably exceeding the commonly estimated mutation rate by a few orders of magnitude. Daniel R. Wilson (1993) used an established population genetics calculation, the Hardy-Weinberg equation, to demonstrate that bipolar disorder, a heritable condition, is 300 times more common than it should be if we assume that natural selection doesn't favor bipolar-related genes. Because the epidemiological characteristics of bipolar disorder and schizophrenia are so similar, the calculation for schizophrenia yields a comparable result.

It is improbable that a mutational hotspot occupying a single section of DNA could explain schizophrenia, due to mounting evidence that numerous genes contribute to schizophrenia.

Schizophrenia and Reduced Fecundity

Fecundity and fertility are sometimes used interchangeably, but these terms are not equivalent. Fertility refers to the biological ability to produce offspring. Fecundity, on the other hand, represents the number of descendants one produces, taking into account all factors including biological ability (fertility) and social opportunity. Patients with schizophrenia tend to have fewer intimate relationships, and so fewer opportunities to reproduce. It is therefore fecundity that is compromised in schizophrenia, not fertility.

Reduced fecundity, particularly in males with schizophrenia, is a reliable finding in Western societies and is clearly due to fewer conjugal relationships (Haukka et al. 2003; Haverkamp et al. 1982; McCabe et al. 2009; McGrath et al. 1999; Nanko and Moridaira 1993). However, no accurate estimates exist beyond the last half-century and no figures have ever been established for traditional societies. This means that reduced fecundity in schizophrenia could conceivably be a modern cultural artifact.

We can all agree that the fecundity of schizophrenia (or shamanism) in ancient traditional societies could have only followed three possible courses: increased, decreased or neutral. Consistently elevated fecundity relative to the general population would have been unlikely for the simple fact that schizophrenia never grew to be the predominant phenotype. This leaves neutral and reduced fecundity as the most plausible alternatives.

I am inclined to argue that perfectly neutral fecundity is unlikely for such a deviant phenotype, especially one that impedes interpersonal relationships. What are the chances that people with such a distinctive condition have the exact same fecundity as the rest of humankind? There are plenty of noticeable phenotypic traits that may or may not confer reproductive advantage; for example, eye color. Although certain eye colors may be more attractive than others, it is conceivable that these preferences don't translate to any substantial reproductive advantage. Any preferences for eye color may be trumped by preferences for other more important physical or personality attributes. In contrast, schizophrenia changes people in ways that significantly reduce the quality of intimate relationships. Thus, it would seem that the severest forms of schizophrenia would be accompanied by reduced fecundity.

Robert Wright (2009), author of *The Evolution of God*, has suggested that the modern image of the shaman is glorified, and that traditional shamans could have been as crass and opportunistic as anyone else. Given how some contemporary shamans have operated — exchanging sex for their services — such an idea is not out of the question.[1] It is certainly possible that the fecundity gap for schizophrenia (or shamanism) could have been much narrower in traditional societies. For example, Bhatia and others (2004) could not find any reductions in fecundity among a small patient sample from India. This is consistent with other studies that suggest schizophrenia is less socially and occupationally debilitating in less-developed countries. The authors, however, acknowledged that the Indian cultural practice of arranged marriages could have explained this atypical result. In addition, this study was not statistically powered to distinguish the relatively small differences in fecundity that could conceivably exist — small differences that may well be very important on evolutionary timescales.

If one supposes that schizophrenia is heritable and many thousands of years old, its fecundity would have to be almost equal to that of the general population — smaller than a 1 percent difference — otherwise schizophrenia would have disappeared. That is because small differences in fecundity can, over time, translate into huge disparities in the frequency of the corresponding trait. For example, if schizophrenia were accompanied by a mere 1 percent reduction in offspring, the prevalence of schizophrenia would be reduced by about 50 percent after only 600 generations, or 15,000 years (Ridley 1996, 102).[2] In other words, a tiny 1 percent decrease in offspring, relative to the general population, would eventually — over a 15,000-year interval — diminish the frequency of a heritable trait by one half. If we presume that schizophrenia is ancient, heritable, and occurs within a distinct group of people of reduced fecundity, then something does not add up. Such a scenario does in fact turn out to be possible, but only if auxiliary evolutionary mechanisms, like balanced polymorphism, are introduced to countervail the adverse aspects of schizophrenia. We will explore such possibilities in later chapters.

The inherent problem with such debates over fecundity is that they all apply the prevailing *categorical* approach to schizophrenia: schizophrenia is treated as a completely distinct phenotype with a genetic profile independent from that of the rest of humanity. The alternative is a *di-*

---

[1] http://www.slate.com/id/2223786/

[2] A 1 percent reduction in offspring translates to a selection coefficient of 0.01. Assume, for simplicity's sake, a single locus Hardy-Weinberg equilibrium to get the 50 percent reduction in 600 generations.

mensional approach, which views schizophrenia as the extreme end of a continuum which includes more conventional behavior. As we shall discover in later chapters, a simple categorical approach to schizophrenia may not be entirely accurate (Esterberg and Compton 2009; Linscott and van Os 2010). Dimensional models could change our assumptions about the evolutionary dynamics behind schizophrenia.

When looking at complex polygenic traits (e.g., height, weight) on a bell curve, one expects the extremes of phenotypic expression to produce reduced fecundity. In other words, people of average height will have the greatest reproductive advantage by avoiding the worst drawbacks of being too short or too tall. All individuals must possess height or weight, so those bell curves include the entire population. In contrast, the great majority of people do not have schizophrenia. Milder forms of schizophrenia (i.e., schizotypal or schizoid personality traits) exist, but only in a small minority of people. Therefore, there is only acceptable evidence for a partial continuum.

It is certainly possible that a widespread polygenic trait — for example, creativity — could be surreptitiously connected to schizophrenia. In such a scenario, the most deviant forms — for example, the extremes of creative expression — somehow become transformed into schizophrenia.

Whether schizophrenia is better represented as a distinct category or part of a phenotypic continuum has become a contestable point among psychiatric researchers (Lawrie et al. 2010; Peralta and Cuesta 2007; Stip and Letourneau 2009; van Os and Verdoux 2003). It is wholly possible that both camps have merit and that some sort of hybrid model is in operation — one that incorporates both categorical and dimensional approaches to schizophrenia.

The Geographical and Temporal Stability of Schizophrenia

In contemporary Western societies about 1 percent of the population has schizophrenia (Hafner and an der Heiden, 1997). Because Western nations are to some extent culturally homogeneous the presence of schizophrenia could be linked to some ever-present environmental factor. Thus, one burning question is whether schizophrenia exists in every corner of the world. If a population existed without schizophrenia, one could search for possible environmental or genetic differences. In the absence of such an ideal population, another question worth exploring is whether the prevalence of schizophrenia varies — even a little — throughout the world. It would also be helpful to know whether the prevalence of schizophrenia has increased, decreased or remained unchanged throughout history.

A review of 70 prevalence studies confirmed that the lifetime prevalence for schizophrenia is approximately 1 percent throughout the world

(Torrey 1987). There was some variability between populations; from 0.03 percent in an Amish population to 1.7 percent in an isolated population in Northern Sweden. Such differences are not unexpected when one factors in random statistical deviations caused by studying smaller populations, cultural differences in diagnosing schizophrenia, cultural differences in the presentation of schizophrenia, the wide variety of epidemiological techniques used to identify schizophrenia (some looked at old hospital records while others interviewed people in the general population) and different measures of prevalence (point, yearly, lifetime and morbid lifetime risk).

The landmark World Health Organization Ten Country Study attempted to control for some of these factors (Jablensky et al. 1992). The use of uniform study criteria and consistent interview techniques considerably reduced the variability in prevalence rates across the ten chosen countries. The authors concluded that "schizophrenic illnesses are ubiquitous, appear with similar incidence in different cultures and have clinical features that are more remarkable by their similarity across cultures than their difference." There was only a two-fold difference between the highest and lowest incidence rates. (And a three-fold difference when applying broader criteria for schizophrenia.)

A few researchers seem convinced that ascertainment, or sampling, bias alone cannot sufficiently explain differences in prevalence rates between jurisdictions (Messias et al. 2007; McGrath 2006, 2005; Saha et al. 2005; McGrath et al. 2008; Goldner et al. 2002). Challenging assumptions is always valuable, but there appears to be no evidence to support the notion of genuine differences in schizophrenia rates. Psychiatric epidemiology studies are notoriously complex and erratic results are not uncommon. If one factors in the expected deficiencies in inter-rater reliability, especially between disparate cultures, the global schizophrenia prevalence figures seem remarkably uniform (Grove 1987; Mojtabai and Nicholson 1995; Hickling et al. 1999). That is not to say that schizophrenia must have precisely identical prevalence rates in every part of the world, but as of yet, there is little evidence to suggest that there are any meaningful irregularities.

Incidence is the measure of new cases of a medical disease within a given timeframe, usually one year. It is a superior measure, compared to prevalence, for ascertaining short-term historical trends. A number of studies have attempted to determine whether the incidence of schizophrenia has increased, decreased, or remained fixed over the last few decades. Several studies suggest declines in incidence throughout various regions of the Western world (Finland, Australia, Denmark, England and Wales, Italy, Ireland, New Zealand, and Scotland) whereas a smaller group of studies conducted in other regions (Croatia, Netherlands and parts of

England) show no notable differences (Brewin et al. 1997; Der et al. 1990; Suvisaari et al. 1999; Waddington and Youssef 1994). A number of possible confounding variables exist that could significantly alter incidence rates, including deinstitutionalization, stricter diagnostic criteria and earlier treatment of psychosis (Kendell et al. 1993; Stromgren 1987). When one factors in all the conflicting studies and all their possible confounding variables, there simply isn't any appreciable evidence supporting significant deviations in schizophrenia incidence over the last century (Jablensky 2003).

Other Major Epidemiological Factors

The characteristic age of onset of schizophrenia for males is 15–25 (Hafner 2003). Women usually show symptoms, on average, a few years later (Hafner 2003). Symptoms of schizophrenia generally present before age 50. New cases of schizophrenia are rare in the elderly and equally infrequent in preadolescent children.

It has been conventionally accepted that sex ratios of schizophrenia are approximately equal (Jablensky 2003; Hafner and an der Heiden 1997, 2003; Jablensky and Cole 1997). To my knowledge, no studies have ever shown increased rates in females compared to males. However, a few studies have found higher rates for males. In fact, two recent studies have yielded very similar male to female ratios of approximately 1.4, a difference in prevalence rate that can not be explained away by numeric deviations produced by the earlier onset of schizophrenia in males (McGrath 2006; Aleman et al. 2003). These supposed differences in sex ratios of schizophrenia are in accordance with my own clinical experience. If gender disparity in schizophrenia exists, it could be due to inherent biological sex differences or the possibility that symptoms of schizophrenia are more disadvantageous to the male social role (Nicole et al. 1992; Tamminga 1997; Leung and Chue 2000).

Springtime births may be associated with slightly increased rates of schizophrenia: 5–8 percent at most. The southern hemisphere spring (September) has also been weakly associated with increased schizophrenia (Cannon et al. 2003, 75). Urban birth may also slightly increase the risk of psychosis in Western societies (Boydell and Murray 2003, 55).

How Old is Schizophrenia?

There are several lines of evidence suggesting that schizophrenia is a very old phenomenon, but tracing the history of schizophrenia has its hurdles. The following section will discuss obstacles to identifying medical ailments in the historical record, historical examples of psychosis (schizophrenia and bipolar disorder), and finally, estimating the evolutionary age of schizophrenia.

Explaining the Shortage of Schizophrenia Cases in the Historical Record

Jeste and others (1985) were the first to put forward a list of possible reasons for the shortage of unequivocal cases of schizophrenia in the historical record. They recognized six major hurdles impeding the identification of schizophrenia in ancient manuscripts. I will follow their general outline and expand upon some of their points.

First, ancient physicians viewed medical symptoms as a guide to overall health rather than thinking in terms of precise diagnoses. The zealous pigeonholing of signs and symptoms into discrete disease categories is routine for twentieth century physicians, but in ancient times, without treatments and a sense of pathophysiology, pinpointing a diagnosis would have been futile. For example, Jeste and others point out that cholelithiasis (gallstones) was rarely described before the fifteenth century, but became increasingly recognized after autopsies began to be performed. There is no doubt that ancient physicians were astute observers, but certain subtle symptoms and signs linked to schizophrenia would have been elusive. For example, negative symptoms of schizophrenia such as poor motivation or affective flattening are not easily discernable, even for psychiatrists with contemporary psychiatric training. Recall that the pioneers of modern psychiatry like Bleuler and Kraepelin were among the first physicians ever exposed to large numbers of mentally ill patients, which allowed them to discern subtle behavioral patterns.

A second reason for the shortage of descriptions of severe mental illness in ancient medical records could be a lack of interest in chronic incurable illnesses. Ancient physicians were in the curing business — their livelihoods depended on efficacious interventions. There is good evidence that they did not involve themselves with intractable behavioral disturbances or taking care of the dying. (The notion of formal medical palliative care is a fairly recent phenomenon.) Historian Keith Thomas writes of the climate of medicine in Britain during the early Renaissance (1971, 15):

> Even less could be done for sufferers of mental illness. Contemporary medical therapy was primarily addressed to the ailments of the body. "For diseases of the mind" wrote Robert Burton,[3] "we take no notice of them". Raving psychotics were

---

[3] Robert Burton was a seventeenth-century philosopher who in 1621 published *The Anatomy of Melancholy*, one of the first European books on mental illness.

locked up by their relatives, kept under guard by parish officers, or sent to houses of correction.

The third reason for the shortage of reported schizophrenia cases in ancient times may have been that milder forms of mental illness often went unrecognized by families because such behaviors did not result in any social disturbance or personal distress. This situation is not that old; early in my career, I met several families of psychotic patients that had additional members whose thinking was clearly disordered. The families typically framed those members' nonsensical language as poetic, mystical or esoteric. Compared to previous generations, contemporary society is much more sophisticated and self-aware about psychological matters — as well as being less tolerant of peculiar behavior.

The fourth reason for the shortage of schizophrenia cases in the historical record relates to the antiquated idea that spiritual delusions are religious quandaries, not medical disturbances. This was particularly true in the Middle Ages during the ascendancy of European Christianity, when the church was very much intertwined in the daily life of the community. There is agreement among medical historians that, prior to modern times, the greatest record of unrecognized schizophrenia was found in reports of experiences interpreted as supernatural or spiritual (Turner 1992). For example, Kroll and Bachrach (1982) state, "In Western Europe, from 500 to 1500 AD, people who heard voices or saw visions considered themselves, and were considered by their contemporaries, to have had actual perceptual experiences of either divine or satanic inspiration. They were not considered mad and were not dealt with as such." In a similar vein, the psychiatric historian Chris Philo (2003) concludes, "...the difficulty of providing a 'rational account' of mental disorder — a difficulty that is hardly any less today — must on occasion have led pre-scientific peoples, both pagan and Christian, to view madness as the doings of devils, demons and other unpleasant entities" (p. 88).

Keith Thomas's historical analysis of spiritual, religious and magical beliefs pertaining to sixteenth- and seventeenth-century England reveals several examples of apparent spiritual delusions that were overlooked by medical practitioners (1992, p. 157):

> The reign of Elizabeth produced a small army of pseudo-Messiahs. In London in April 1561 John Moore was whipped and imprisoned for saying he was Christ. A month later a 'stranger' was set in the stocks for claiming to be the Lord of Lords and King of Kings. In the following year Elizeus Hall, a draper, was arrested and interrogated by the Bishop of London for assuming the title of Eli, the carpenter's son. He confessed

to having had visions in which he had been selected as a messenger from God to the Queen and privileged with a two-day visit to Heaven and Hell.

These vignettes are just a few of the innumerable accounts of religious delusions throughout history. It is doubtful that these stories were fraudulent, because of the risk of serious castigation. According to Thomas, educated people often sensed that such beliefs represented some sort of malady and for the sake of humaneness, tried to ignore such "brainsick" or "frantic" individuals (p. 157). By the seventeenth century, intellectuals such as Thomas Hobbes and John Locke were unequivocal in their denunciation of popular superstitions. Locke, for example, argued that belief in witches and goblins was silly, and that witches weren't diabolical but delusional (Porter 2002, 60). In 1786, a psychotic woman attempted to stab King George III with a cake knife. King George, perhaps because he himself struggled with periods of psychosis, asked that the woman receive sympathetic care at London's Bethlem Royal Hospital (Arnold 2008).

The fifth reason for the shortage of schizophrenia cases in the historical record relates to confusing or misunderstood terminology in ancient texts. Terms such as phrenitis (delirium), mania and melancholy had slightly different meanings depending on the author and epoch.

Perhaps the greatest confusion lies in the fact that bipolar disorder and schizophrenia were rarely distinguished prior to the twentieth century. Because every symptom of schizophrenic psychosis could be a symptom of manic psychosis, but not the other way around, any comprehensive description of psychosis that does not differentiate between these two conditions will appear to be describing mania. The five chief symptoms of schizophrenia as marked out by DSM-IV are:
   i.   delusions,
   ii. hallucinations,
   iii. disorganized speech,
   iv. grossly disorganized or catatonic behavior, and
   v.   negative symptoms.

The seven major symptoms for mania as outlined by DSM-IV are:
   i.   grandiosity,
   ii. insomnia,
   iii. talkativeness, more so than usual
   iv. flight of ideas,
   v.   distractibility,
   vi. increased goal-directed activities or psychomotor agitation, and
   vii. impulsivity in usually pleasurable activities.

Setting aside negative symptoms, which are very subtle and not dependably recognized until the early twentieth century, if you add the four remaining psychotic symptoms of schizophrenia to the seven manic symptoms, the resulting list appears to be a description of manic psychosis, and not schizophrenia.

Let's take, for example, the medical writings of the Greeks. Greek physicians generally classified emotional-behavioral disturbances into three basic categories: melancholia (sometimes with psychosis), mania and delirium. One can imagine that most cases of schizophrenia would have been pigeonholed as either melancholia — psychotic depression — or mania — manic psychosis.

Another example of ancient diagnostic ambiguity can be found in the 2000-year-old Chinese text *The Yellow Emperor's Classic of Internal Medicine* which distinguishes delirium-like states from "craziness" (*kuang*) but does not further subdivide insanity (Lam and Berrios 1992). Descriptions of *kuang* appear to be more consistent with bipolar disorder precisely because insanity is not divided into Kraepelin's two categories. According to Lam and Berrios, *The Yellow Emperor's Classic of Internal Medicine* seems to describe both schizophrenia and mania under the singular category of *kuang*:

- Those with the illness of craziness (kuang) have visions, hear voices, and scream out...
- Their speech is irrelevant, is easily frightened, smiles inappropriately, sings and exhibits abnormal behavior all day; this might result from extreme fear...
- When the attack of craziness occurs, the patient does not want to sleep, does not feel hungry, and considers himself to be someone out of the ordinary, exceptionally clever or noble. There are also manifestations of abnormal thinking.
- The crazy patient eats a lot, sees spirits and devils, smiles to himself without communication to others; this may result from excessive happiness disturbing the spirit of the person...
- It is said that some patients shut off their doors and windows, prefer solitude...

(It should be noted that inappropriate smiling is characteristic of schizophrenia and much less commonly observed in bipolar disorder. Manic patients tend to smile appropriately in the context of their delusions.) The preference for solitude could describe schizophrenia, but also perhaps psychotic depression. Although auditory hallucinations and spiritual visions can be part of a manic psychosis, they are more commonly associated with schizophrenia.

In a similar vein, a 2000-year-old text from India, the *Charaka Samhita*, describes a variety of conditions that seem to include both schizophrenia and bipolar disorder (Bhugra 1992). For example, Bhugra's translation of the *Charaka Samhita* reveals eight forms of "possession states" — psychological disturbances that appear to represent psychotic illnesses. Six subtypes resemble various aspects of manic or depressive psychosis. The other two subtypes could very well describe schizophrenia. The first: "[possession by] Gods. Gentle-looking, dignified, indomitable, placid, disinclined to sleep and eat, with scanty excretions, pleasant body-odour, and face radiant like a fully blossomed lotus flower" and the second: "Tend to affect someone devoted to solitude and frequent bathing, and well-versed in law-books and scriptures."

The sixth and last reason put forward by Jeste and others for schizophrenia's scarcity in the historical record may be connected to societal transformations — cultural changes that alter psychological complexes, which in turn vary the expression of mental conditions (1985). For example, catatonic psychosis, an exceptionally dramatic symptom of schizophrenia, has been in decline over the twentieth century and the decline seems to have continued since the publication of the article twenty-five years ago. No one knows why this symptom is no longer regularly seen — although expeditious treatments and greater psychological awareness in the general population could be factors.

Examples of Psychosis Before the Era of Industrialization

Cases of schizophrenia in the remote historical record can always be disputed. This is because any secondary cause of psychosis, like a brain tumor or seizure disorder, can never be completely ruled out. A bout of psychosis occurring in an elderly person, for example, could conceivably be dementia. The concomitant presence of a serious medical condition (capable of causing delirium) can also call a diagnosis of schizophrenia into question. Although depressive symptoms or extreme motor agitation can accompany schizophrenia, these symptoms may indicate bipolar disorder or psychotic depression. Furthermore, drunkenness could always be an explanation for peculiar behavior; historians believe that through much of European history, particularly during the Middle Ages, diluted wine and beer were common sources of hydration, because water was generally not considered potable (Vallee 1997, 125). With all these caveats pertaining to the historical record, a slam-dunk diagnosis of schizophrenia is unrealistic.

Descriptions of human behavior tended to be succinct in older historical records. Before the Renaissance, it was unusual for writers to extensively document the lives of ordinary people. For example, no thorough accounts survive of the hundreds of insane people who first resided in Bethlem Royal Hospital. When references were made to unusual behaviors, they tended to be perfunctory. A typical example can be found in Mas's study (1994, 476) of psychiatry in the Spanish island of Majorca during the Middle Ages: "And in 1382 Coloma, the wife of Joan Cerda, claimed that for some time her husband 'has been out of his mind and has become mad' and 'asked for help in looking after him'." This is, unfortunately, the extent of the information provided.

Even with such obstacles, Jeste and others describe a number of cases throughout history that could very well be schizophrenia (1985). One of the oldest descriptions concerns Cleomenes I, King of Sparta, during the fifth century BC. According to the renowned Greek historian Herodotus, "Now, Cleomenes had been more or less insane before, and no sooner had he got back to Sparta than he succumbed to an illness whose symptoms were that he used to poke his staff into the face of any Spartiate he met. Faced with this behavior and his derangement, his relatives put him in the stocks" (Herodotus 1998). Soon after this incident, Cleomenes died of self-inflicted knife wounds.

Chris Philo describes two cases dating back to about 700 AD which could be schizophrenia. Quoting various antique records, Philo states: "The first involved a young man of noble stock called Hwaetred who was attacked by an 'evil spirit' and became 'affected with so great a madness that he tore his own limbs with wood and iron and with nails and his teeth', and who was generally so violent that those men who sought to subdue him risked serious injury and even death". The second case from the same period describes a young man tormented by an "unclean spirit" and his "powers of speech, discussion and understanding failed him entirely." Months later, his "madness had disappeared" (2003, 108).

Perhaps the earliest up-to-standard description of schizophrenia madness concerns seventeenth-century Bavarian painter Christoph Haizmann. Haizmann experienced apparitions of Jesus and the Virgin Mary, and visions of being taken to hell. He frequently heard the voice of the devil, and felt repeatedly tormented by the devil over a number of years — an experience commonly observed in modern-day schizophrenia. Otsuka and others (2004) argue that bizarre delusions are an especially characteristic symptom of schizophrenia, uncommonly observed in other psychotic conditions. They argue that the prominence of Haizmann's religious delusions were clearly abnormal — even for seventeenth-century European culture.

Some psychiatric researchers remain unconvinced that schizophrenia existed to any appreciable extent before the 1800s (Torrey and Miller 2001). In a recent systematic review of Greek and Roman literature, Evans and others (2003) concluded there was no concrete evidence for schizophrenia in ancient history. However, this study applied restrictive modern criteria, including requiring evidence for chronicity of illness — arguably a very high bar to surpass without the aid of an hour-long structured DSM-III interview. The authors claim that many other psychiatric disorders, in contrast, are found more abundantly in the historical record. However, in our own study exploring the history of obsessive-compulsive disorder, a diagnosis whose prevalence rate is comparable to that of schizophrenia, we could not find any unequivocal cases in the remote historical record (Polimeni et al. 2005). Similarly, posttraumatic stress disorder, with modern prevalence rates around 5 to 8 percent, is rarely described, even superficially, in the writings of antiquity.

The Rise of Schizophrenia

The twentieth-century French philosopher Michel Foucault was among the first to address the apparent rise in mental illness over the last few centuries (Foucault 1988). Foucault believed that mentally ill people had been slotted into the cultural niche left vacant by the decline of leprosy in Europe. The gradual displacement of superstitious thinking in the post-Renaissance Age of Reason made the mentally ill yet more conspicuous. Foucault felt that burgeoning economic pressures created greater scrutiny towards capitalistically unproductive members of the community, including the mentally ill. For Foucault, the fifteenth-century literary allegory of the "ship of fools," a boat loaded with the insane and sent off to a faraway jurisdiction, described the genuine desire of certain European communities to rid themselves of undesirable people. Among medical historians, there seems to be agreement with Foucault's general claim (but not necessarily every detail) that people with insanity had particular trouble adapting to the post-Renaissance world.

Kroll's (1973) historical analysis of mental illness during the Middle Ages pointed to urbanization as a primary reason for the greater visibility of psychiatric conditions. He concluded, "As the population of Europe began to increase in the 11th century, and as the towns became centers of commercial and industrial growth, the care of the mentally ill gradually shifted from a family to a community problem."

However, another study suggests that the increase in the number of mentally ill is due to industrialization rather than urbanization per se. Chris Philo (2003) recently completed an exhaustive analysis of the institutions that housed the "insane" from medieval times to the nineteenth century in England and Wales. He meticulously documented the gradual and steady rise in the number of mentally ill people housed in British institutions since the Middle Ages. Philo's analysis reveals several reasons why mental illness has become more prominent in recent centuries: secularization, specialization of labor, and increased government accountability. These three factors are perhaps better captured by the single term *industrialization*.

The rise of industrialization during the eighteenth century marks the initial escalation in cases of schizophrenia. This is not likely to be a coincidence, because the societal changes that accompany industrialization are especially problematic for patients with schizophrenia. Agricultural work, although physically demanding, is better structured to tolerate fluctuations in output. In contrast, industrialization places greater emphasis on unfailing motivation and reliability. To be, as Karl Marx described, "an appendage of the machine" requires that workers reliably attend their posts at specific times and maintain consistent enthusiasm throughout the day — a task that is particularly difficult for those with schizophrenia. Further, in the post-industrial era an inconsistent resume has always made it difficult to get hired. Because 80 to 90 percent of patients with schizophrenia are not employed, occupational impairment has consequently become a diagnostic criterion for the disorder — arguably a form of circular thinking.

In early seventeenth century England, only 8 percent of people lived in communities with a population of 5000 or more. Next door in France, the figure was estimated at 8.7 percent. By 1800, these numbers rose to 27.5 percent in England and 11.1 percent in France (Harris 2003, 226). By 1900, over half of British people were living in cities (Stearns 1994, 780). Other Western European nations followed similar trends, experiencing population shifts that were a direct result of industrialization. And no matter the jurisdiction, the rise of schizophrenia followed the same trajectory. In Russia, for example, where industrialization lagged behind Western Europe, there were hardly any dedicated asylums before 1850 (Porter 2002, 94). Similarly, Portugal only had two insane asylums at the close of the nineteenth century (Porter 2002, 94).

In addition to the sociocultural changes produced by industrialization, Philo suggests other practical explanations for the relative paucity of reports of mentally ill persons in British historical records before the eighteenth century. First, most places housing the severely mentally ill were small, usually with only two or three residents. Even London's notorious Bethlem Royal Hospital (from which we get the word *bedlam*) contained merely forty-eight inpatients in 1659 (Philo 2003, 126). Second, the unofficial residences housing the insane tended to be secretive in order to protect families from stigma. Third, informal arrangements tended to generate little paperwork, especially during pre-bureaucratic times.

Circumstances began to change in England by 1774 when private madhouses were required to be licensed — a watershed moment for historical records. Philo observes, "Reliable data on the numbers, sizes and locations of madhouses is nonetheless impossible to find before 1774, as much of the information from earlier years is haphazard and anecdotal" (2003).

It thus appears that over the last two centuries, a number of significant cultural factors converged, causing schizophrenia to gain its present visibility. The most significant cultural change may be industrialization, with its specialization of labor, detailed record keeping and greater government accountability. Secularization, which may be indirectly related to industrialization, further alienated psychotic persons. Such societal shifts made psychotic ailments more likely to be social problems, which explains the spike in cases seen in developing countries. As nations became wealthier, their formidable bureaucracies spawned substantial healthcare institutions such as nursing homes, substance-abuse treatment centers and mental institutions: "If you build it, they will come." These establishments, in turn, had the unintended effect of reinforcing stereotypes and oversimplifying how we characterize patients and their medical conditions.

Dating Schizophrenia

Dating schizophrenia precisely is presently not possible, although it may be, at such as time as the specific operative genes become known. By estimating the rate of mutational changes in DNA, molecular dating techniques can approximate the timing of evolutionary divergence between various genotypes. This, of course, is not yet possible with classic psychiatric disorders since the genes involved have not been identified, but it may be doable in the not-so-distant future.

A minimum estimate of schizophrenia's evolutionary age is, however, feasible. There is reasonable evidence to suggest that schizophrenia is at least 50,000 years old (Polimeni and Reiss 2002). This estimate was achieved by putting four assumptions together: that schizophrenia is heritable, that schizophrenia is present in Australian Aborigines, that Australian Aborigines colonized Australia about 50,000 years ago, and that Australian Aborigines have been (relatively) genetically isolated since migrating onto the continent.

We have already concluded that schizophrenia is a highly genetic condition, with some heritability estimates as high as 85 percent. This leaves three remaining assumptions to examine.

Australian psychiatrists have routinely reported schizophrenia in Australian Aborigines (Mowry et al. 1994). In fact, the specific incidence and prevalence of schizophrenia in the Australian Aboriginal population and in other nearby remote populations such as that of Micronesia appear to be comparable to the rest of the world (Mowry 1994; Myles-Worsley et al. 1999; Torrey 1987; Torrey et al. 1974; Waldo 1999; Jones and Horne 1973; Murphy and Raman 1971). The precise symptom profile, however, may be slightly different from non-Aboriginal patients — presumably due to cultural factors. Jones and Horne, for example, observed three cases of schizophrenia in a sample of 959 Australian Western Desert and Kimberley Aborigines. Three other cases were "possession states," which could also have been schizophrenia. (They estimated that about 90 percent of their population sample would be considered full-blooded Aborigines.) Moreover, Jones and Horne implied that their prevalence figures likely underestimated the true frequency of schizophrenia due to the formidable difficulties in recognizing schizophrenia in disparate cultures.

H. D. Eastwell made similar observations of schizophrenia among the Australian Aborigines of the Arnhem region (1982). Acute psychosis frequently manifested as a conviction that one was a victim of sorcery. Individuals with chronic and severe psychotic symptoms were usually exempt from normal tribal activities. Eastwell observed, "Reciprocal returns of food or money are not expected from them..." It is also interesting that in certain Aboriginal dialects, psychosis is indicated by the word "deaf" — the implication being that the "ears are blocked by the noxious spirit in their heads."

Various evidence suggests that the ancestors of modern Australian aborigines settled Australia about 50,000 years ago (Brown 1997). The majority of archeological sites have been carbon-dated to before 40,000 years ago. Archeological evidence points to relative cultural isolation with at most minor contact with outsiders on the northern coast of Australia (Kingdon 1993; Brown 1997). The first Western explorers that encountered Australian Aboriginal tribes observed stone-age technology, which is in keeping with theories of relative cultural isolation.

A recent genetic study of Australian Aborigines estimated divergence from other Eurasians at about 50,000 years ago. Genetic testing, specifically of mitochondrial and Y chromosomal DNA, also suggests considerable genetic isolation since the initial colonization (Hudjashov et al. 2007). This means that schizophrenia probably existed before the establishment of the oldest isolated racial enclaves 40,000 to 50,000 years ago. It is therefore certainly possible that schizophrenia existed prior to *Homo sapiens*' exodus from Africa about 50,000 to 60,000 years ago.

Conclusion

In this chapter, we have considered the main parameters of the schizophrenia paradox. Schizophrenia appears to be a heritable condition whose incidence greatly exceeds normal mutation rates. Schizophrenia has probably existed for a very long time, although its precise form, frequency and severity may have varied. Contemporary fecundity rates associated with schizophrenia lag behind those of the general population. The reason people with schizophrenia have fewer progeny appears to be related to social impairment. Although it may be supposed that some degree of social impairment and resulting reduced fecundity was also present in pre-historic times, there is no way to verify this.

Although a number of minor doubts prevent us from fully embracing the schizophrenia paradox, there is nonetheless substantial evidence for its validity. At least, it should prompt us to search for countervailing evolutionary benefits related to schizophrenia. In some ways, the schizophrenia paradox can be viewed as a warning light: it does not necessarily prove that schizophrenia has evolutionary benefits, but it certainly alerts us to that possibility.

– 4 –
Is Schizophrenia a Disease?

We have reviewed the overt clinical manifestations of schizophrenia: delusions, hallucinations and disjointed thinking. Now we will look beyond schizophrenia's behavioral manifestations and examine everything else that is known about it. As we appraise the most up-to-date pathophysiological findings, I will challenge the belief that schizophrenia is strictly a disease. The greatest obstacle for the shamanism theory may be the fact that schizophrenia resembles conventional medical diseases in certain ways. However, it is also true that schizophrenia lacks many of the characteristics observed in almost every other medical disease. Under closer scrutiny, it will become evident that schizophrenia is a unique biological phenomenon.

Our closer look at schizophrenia will begin with the neurobiology of schizophrenia. Beneath the surface of observable clinical symptoms lie a number of subtle biological differences between people with schizophrenia and healthy people. Schizophrenia has been examined from a variety of angles: neurohistological, neuromolecular, neuroanatomical, neurophysiological, neurocognitive and neurobehavioral. Understanding the differences between people with schizophrenia and healthy people promises the possibility of characterizing the underlying machinations of schizophrenia.

Second, we will review the existing literature concerning the genetics of schizophrenia. Although the Holy Grail — a single gene responsible for schizophrenia — is probably a pipe dream, a hazy outline of the hereditary factors behind schizophrenia is beginning to emerge. Medical genetics is a rapidly developing field and we may be on the brink of some enlightening discoveries. Since DNA often reveals its own evolutionary history, we may be very close to confirming (or discarding) certain evolutionary theories related to schizophrenia.

Third, we will examine the nature of medical diseases and how schizophrenia fits into the family of neuropsychiatric disorders. Because biological organisms tend to become disabled in predictable ways, it is worth exploring commonalities intrinsic to all medical diseases. There are a million ways for a car to break down, but in practice, the majority of vehicle failures are the result of one of a short list of mechanical flaws. Most of these glitches can be traced back to a specific design weakness or manufacturing issue. Similarly, most medical diseases are inextricably linked to their evolutionary history. It will be important to consider how schizophrenia fits alongside other human diseases — especially neuropsychiatric defects.

Before closing the chapter, we will explore the kinds of psychosis associated with medical conditions. Delusions and hallucinations can be caused by medical conditions such as dementia, epilepsy, or delirium caused by intoxication or fever. This means that psychosis is not exclusively associated with classic psychiatric disorders like schizophrenia and bipolar disorder. Although the presence of psychotic symptoms in a genuine medical disorder is a compelling reason to frame schizophrenia as a medical disease, it does not seal the deal. We will examine the differences that exist between medically induced psychosis and schizophrenia — disparities that may give weight to the shamanism theory of schizophrenia.

Neurobiology of Schizophrenia

This section will explore major findings about schizophrenia from various branches of neuroscience.

Neuromolecular Findings

A number of synaptic neuroreceptors are likely to play a role in the pathophysiology of schizophrenia. Three types of neuroreceptors have received the greatest attention from schizophrenia researchers: dopamine, serotonin and glutamate.

For decades, the dopamine theory of schizophrenia has been one of the most popular explanations of the principal pathophysiology of schizophrenia. It postulates that psychotic symptoms in schizophrenia are the result of excessive dopaminergic activity in certain parts of the brain, particularly the mesolimbic and mesocortical dopaminergic neural pathways. The best supporting evidence is that almost every major antipsychotic medication blocks dopamine D_2 receptors, a subtype dopamine receptor (Schatzberg et al. 2005). Moreover, a number of substances that are dopamine agonists, such as cocaine and amphetamines, have a propensity to produce psychosis. It is certainly possible that differences in dopamine receptor densities are linked to a propensity for psychosis (Miyamoto et al. 2003).

The serotoninergic system may also play a role in the generation of schizophrenia symptoms. Most of the newer (atypical) antipsychotics demonstrate substantial effects upon the serotonergic system, specifically the 5-HT 2A and 5-HT 1A receptors, as well as on dopamine receptors. A number of hallucinogens such as LSD seem to preferentially affect the serotoninergic system. This could be compatible with the dopamine hypothesis, because dopaminergic and serotoninergic pathways are known to be able to modify each other (Miyamoto et al. 2003).

In some parts of the brain, glutamate pathways are similarly intertwined with dopaminergic systems, so it's not entirely surprising that the neurotransmitter glutamate, through its signaling effects upon NMDA (N-methyl-D-aspartate) receptors, may have something to do with schizophrenia. In fact ketamine and PCP (phencyclidine), both known antagonists of NMDA receptors, are capable of precipitating psychotic experiences in healthy people. Some speculate that diminished glutamatergic activity in certain cortical regions contributes to the symptoms of schizophrenia (Leonard 1997).

Because of the prospect of establishing an entirely novel family of unique schizophrenia medications, glutamate receptors have recently received increased pharmaceutical interest. Other neurotransmitter systems such as GABA (gamma-Aminobutyric acid) and the alpha-7 nicotinic receptor are also being studied for their possible role in schizophrenia (Keshavan et al. 2008).

For several decades, schizophrenia has been intensively studied at the neuromolecular level; however, no unique pathological mechanism has ever been identified. At most, there is evidence for altered receptor densities in certain parts of the brain.

Neurohistological Findings

The field of neurohistology, the study of the minute structures of the brain, has also come up short in the search for specific pathological markers for schizophrenia. As long as a hundred years ago the renowned neurohistologist Alois Alzheimer placed samples of schizophrenic brains under the microscope and could not find any substantive differences between them and non-schizophrenic brains. Gliosis, a form of cellular damage widely recognized as a marker of degenerative neuropathology, is conspicuously absent. Because of such unsuccessful searches schizophrenia has been called "the graveyard of neuropathologists" (Harrison 1999).

An expert review paper examining microscopic neuropathology in schizophrenia emphasized that the majority of positive neurohistological findings to date have not been satisfactorily replicated (Harrison 1999). This is an important point: schizophrenia research has been characterized by novel findings that, in due course, do not pan out. A perusal of any pile of psychiatric journals from previous decades will reveal countless papers touting promising results in discriminating between people with schizophrenia and healthy individuals — only later to report that the results could not be replicated.

Notwithstanding inconsistencies in replication, it is unfair to entirely dismiss the field of neurohistological research in schizophrenia. There is some evidence pointing to small differences in neuronal densities in specific brain areas. For example, the density of chandelier neuron axon terminals may be slightly reduced in the dorsolateral prefrontal cortex in people with schizophrenia (Konopaske et al. 2006; Pierri et al. 1999). There is also some preliminary evidence for cyto-architectural abnormalities in such parts of the brain as the entorhinal cortex or hippocampus (Keshavan et al. 2008). But if such neuronal disarray does exist, it is neither grossly obvious nor ubiquitous. In fact, these subtle neurohistological differences are usually only detected by statistically analyzing large numbers of microscopic samples (Iritani 2007).

Finally, a few researchers have put forward viral theories of schizophrenia, proposing that an infectious agent such as *Toxoplasma gondii* or cytomegalovirus is responsible for schizophrenia in a minority of cases (Yolken and Torrey 2008). The evidence, for the most part, is weak and circumstantial. Furthermore, it does not appear to explain the majority of cases of schizophrenia. *Toxoplasma gondii*, for example, shows extreme variability in infectious rates (seroconversion rates), from 100 percent in certain tropical areas to almost zero in a population of Alaskan Inuits (Ledgerwood et al. 2003). These wide-ranging results cannot be reconciled with the almost-uniform global prevalence of schizophrenia.

Neuroimaging Findings

An assortment of medical imaging techniques are used to derive visual representations of brain anatomy or function, most commonly computed topography (CT) scans, static magnetic resonance imaging (MRI), diffusion tensor imaging (DTI), magnetic resonance spectroscopy (MRS), positron emission topography (PET) and functional magnetic resonance imaging (fMRI). In psychiatric research these imaging techniques are primarily used to search for differences between the neuroanatomy of people with target disorders and of healthy control subjects.

Pneumoencephalography is a technique invented in the early twentieth century which involved draining cerebrospinal fluid and injecting air into the resulting cavity. A medical X-ray then revealed the dimensions of the cerebral ventricles. It was the first neuroimaging technique to demonstrate differences in the neuroanatomy of patients with schizophrenia. One 1935 medical study examined seventy-one patients with schizophrenia, revealing many cases of cerebral atrophy and enlarged ventricles (Moore et al. 1935). This study antedated the widespread use of comparison control subjects or statistical analysis, so by today's standards such results could only be considered informal observations. Even so, the 1935 conclusions turned out to be consistent with modern studies. Pneumoencephalography was a painful procedure and has since been supplanted by less invasive neuroimaging methods.

The advent of non-invasive neuroimaging methods like CT scans and MRIs was a boon for biologically based psychiatric research. Since the 1970s, hundreds of studies have tried to discern differences in neuroanatomical structure between people with schizophrenia and healthy controls. In general, a 2–3 percent decrease in whole brain volumes, particularly in gray matter, is associated with schizophrenia (Gur et al. 2007; Kumari and Cooke 2006). Certain cerebral structures, such as the prefrontal cortex, superior temporal gyrus, amygdala, and hippocampus seem to be slightly diminished in size (Keshavan et al. 2008). Enlarged cerebral ventricles are also observed, particularly in severe chronic cases (DeLisi et al. 2006). These aberrations tend to be present before the onset of the illness and can worsen around the time of the first psychotic break (Pantelis et al. 2005; Pantelis et al. 2007). Neuroimaging findings in bipolar disorder tend to be analogous but are usually less pronounced (Gur et al. 2007).

Cerebral atrophy in schizophrenia may be exacerbated by factors that are only incidentally associated with the condition. Possible contributors to brain atrophy are poor nutrition (Vogiatzoglou et al. 2008; Chui et al. 2008), alcohol (Nesvag et al. 2007), lack of exercise (Wolf et al, 2011), and antipsychotic medication (Andreone et al. 2007; Ho et al. 2011). To muddy the waters further, there is some preliminary evidence suggesting that smoking (Tregellas et al. 2007), cognitive rehabilitation (Eack et al. 2010) and certain atypical antipsychotics (van Haren et al. 2011) could *preserve* cerebral structures in schizophrenia.

Diffusion tensor imaging is a type of magnetic resonance imaging supposedly able to detect changes in the integrity of cerebral white matter. (It hasn't been established how well DTI reflects white matter integrity.) A few studies have pointed towards subtle disorganization of white matter (i.e., myelinated neuronal tracts) in schizophrenia, particularly in frontal and temporal brain regions (Kubicki et al. 2007; White et al. 2008). Although not every study has yielded positive results, there seems to be enough ancillary evidence to implicate myelin dysfunction as a factor in the pathogenesis of schizophrenia. Based on a number of convergent findings, an oligodendrocyte/myelin dysfunction hypothesis of schizophrenia has been formulated (see review: Hoistad et al. 2009). Preliminary findings suggest decreased numbers of oligodendrocytes in the prefrontal cortex in schizophrenia patients (Schmitt et al. 2009). This dovetails with the discovery that neuregulin (NRG1) — an established candidate gene in schizophrenia — may be integrally involved in oligodendrocyte function (Keshavan et al. 2008).

Magnetic resonance spectroscopy is a non-invasive imaging technique that allows identification and quantification of certain biochemical metabolites in living organisms. Because different atomic nuclei possess their own distinctive magnetic attributes, each metabolite will, in theory, emit a unique magnetic resonance signal. In practice, only a short list of brain metabolites can reliably be identified by this technique (e.g., N-acetylaspartate (NAA), creatine, glutamate, GABA). NAA, for example, is primarily synthesized in neurons and therefore could be a proxy measure for neuronal integrity. A number of studies have demonstrated decreased levels of NAA in the hippocampus and prefrontal cortex of schizophrenia patients (Steen et al. 2005). Similarly, other MRS studies have shown reduced synthesis of certain neuronal membrane phospholipids in the prefrontal cortex of patients with schizophrenia (Gur et al. 2007). Although these results are interesting, it is not yet known what is actually occurring on a cellular level.

The most invaluable type of magnetic resonance imaging may be functional magnetic resonance imaging, a non-invasive way to directly view localized brain activity in real time. It is a fortuitous fact of nature that blood contains substantial amounts of iron — an element that readily interacts with magnetic fields; otherwise functional MRI would not work. Iron in oxygenated hemoglobin finds itself in a slightly different molecular configuration than iron in deoxygenated hemoglobin, which means each molecule has its own distinctive magnetic signature. Magnetic resonance scanners can therefore detect real-time changes in the balance of oxygenated versus deoxygenated blood. From this data, neural activity can be inferred in any part of the brain.

Subjects are typically asked to perform one of three types of tasks inside a MRI magnet: sensory tasks (e.g., watch a bright light), motor tasks (e.g., tap a finger) or simple mental tasks (e.g., two-digit calculation). Because active neurons consume oxygen, the brain areas supporting any given cognitive task will exhibit shifts in the ratio between oxygenated and deoxygenated blood. This, in turn, changes the magnetic field corresponding to those parts of the brain, which then shows up on an fMRI monitor. In practice, these changes can be very difficult to detect because of very low signal-to-noise ratios. Therefore, to tease out localized neural activity, fMRI experiments usually rely on numerous task repetitions, as well as sophisticated statistical analysis.

Schizophrenia has been the psychiatric diagnosis most extensively studied by fMRI technologies. Initially, there were a number of reports of "hypofrontality," meaning that patients with schizophrenia demonstrated less overall neural activity in the frontal lobes during standard mental tasks. However, other studies sometimes showed the complete opposite: "hyperfrontality" in schizophrenia. One widely accepted explanation for this discrepancy was that sometimes patients with schizophrenia were exerting more mental effort during tasks (compared to healthy controls) while at other times disengaging and giving up (Manoach 2003).

fMRI is a novel technology in its early stages of use and as such, awaits refinement. One enhancement has been that most fMRI experiments now try to achieve comparable task performance in every experimental group. Although conflicting results continue to be seen, the usual areas of interest in schizophrenia — such as the dorsolateral prefrontal cortex and the anterior cingulate — frequently show changes in neural activity (Fukuta and Kirino 2004; Gur et al. 2007).

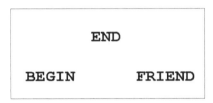

Figure 4.1: An example of an easy semantic triad. BEGIN is the better semantic choice for the target word END. The correct response is according to standardized word association norms (Nelson et al. 1998).

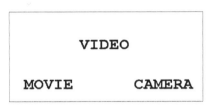

Figure 4.2: An example of a difficult semantic triad. MOVIE is the better semantic choice for the target word VIDEO). The correct response is according to standardized word association norms (Nelson et al. 1998).

A few years ago, Jeff Reiss and I designed an fMRI experiment to examine possible differences in schizophrenic thought processes (Reiss et al. 2007). Because odd or unusual cognitive associations seem to be a fundamental symptom of schizophrenia, we created a word association task especially tailored to be performed inside the magnetic tube of the MRI. Although it is a modest study, it aptly illustrates typical psychiatric fMRI research. Experimental subjects wearing special goggles had a word flashed in front of them. Below this key word trailed two different words potentially related to the key word. Subjects were required to choose which of the two trailing words seemed most semantically related to the key word (Figures 4.1, 4.2). Inside the magnet, subjects performed several dozen trials of our word association task with a different triad of words each time.

Being a pilot study, we only tested eight schizophrenia patients and nine healthy control subjects — a bare minimum to achieve statistical significance. The responses from patients with schizophrenia were a little more inaccurate than the control participants. We also discovered that, relative to healthy control subjects, patients with schizophrenia demonstrated diminished cerebral activity in certain areas of the frontal lobes (middle frontal gyrus). However, we also found that patients with schizophrenia showed enhanced brain activity, compared to healthy control subjects, at several locations within the temporal and frontal lobes. This may not be a spurious finding — another research group conducting a similar word association task obtained very similar results (Kuperberg et al. 2007). Both Kuperberg's team and my own colleagues interpreted the increased localized brain activity as reflecting abnormal neural connectivity in schizophrenia, but this is entirely speculative. In fact, enhanced brain activity is more often than not interpreted in positive terms!

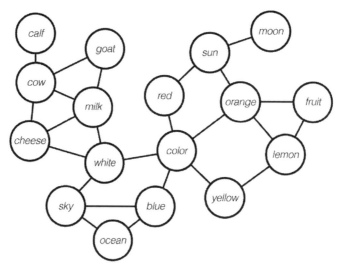

Figure 4.3: A diagram of a schematic semantic memory network. Recreated by permission of the publisher (Minzenberg et al. 2002).

Connectionism is a school of cognitive science that considers how words and ideas could be interconnected and neurophysiologically mapped out in the brain (Figure 4.3). A connectionist model could explain how certain word associations spread neural activity more diffusely than others. The connectionist model has every word represented by its own nodal point inside a three dimensional network of interconnected words. Such strongly associated words as "red," "orange" and "color" are conjectured to be in neuroanatomical proximity, while a faintly associated word like "ocean" may be somewhere downstream. An irrelevant word such as "elevator" would be even farther away from the "red-orange-color" cluster. These associative networks may be altered in schizophrenia, which would explain such variable brain activity. However, "different" does not necessarily mean "worse." Creative people, for example, could conceivably trigger additional words beyond the confines of standard word association networks. Thus, when it comes to semantic word association tasks, a result that deviates from healthy controls does not necessarily mean that something is broken.

Neurocognitive Findings

Schizophrenia is characterized by global cognitive deficits, most obviously executive function, attention, working memory and semantic memory (Sharma and Harvey 2000). Other cognitive domains such as long-term memory, vocabulary and calculation skills are typically less affected. Most cognitive deficits are already present at the time of the first psychotic episode, and the majority of schizophrenia patients do not show further deterioration over the course of their lives (Kurtz 2005; Mesholam-Gately et al. 2009; Moritz et al. 2002; Hoff and Kremen 2003; Russell et al. 1997).

A systematic review analyzing over 200 cognitive studies of schizophrenia supports the conclusion that schizophrenia is characterized by global cognitive deficits (Heinrichs and Zakzanis 1998). The cognitive domains that showed the greatest deficits are verbal memory, full-scale IQ, word fluency tests and continuous performance scores. When one juxtaposes the bell curves representing people with schizophrenia and healthy subjects, the degree of non-overlap is 60–70 percent, a substantial disparity. For most other cognitive tests (e.g., general vocabulary, copying a complex design) a 30–40 percent non-overlap was observed. Moreover, the authors cite a prior study involving monozygotic (identical) twin pairs in which only one twin had schizophrenia; the study found the twins with schizophrenia clearly performed worse than their non-affected twin in the vast majority of cases (Goldberg et al. 1990). Notwithstanding these differences, the authors of this extensive review were still cognizant that "substantial numbers of schizophrenia patients are neuropsychologically normal" (Heinrichs and Zakzanis 1998). In fact, even though most of these studies incorporated significantly ill patients, about 30 percent of schizophrenic participants were able to score above the median of healthy subjects' scores.

The conclusions of this systematic review by Heinrichs and Zakzanis are consistent with my own clinical experience studying humor in people with schizophrenia. Clinical experience has told me that humor recognition seems impaired in schizophrenia, but enhanced in bipolar disorder. Thus, humor appears to be a cognitive domain uniquely suited to studying differences among people with schizophrenia, people with bipolar disorder and healthy control subjects. Humor is a wonderfully complex cognitive ability that seems to rely especially on executive function, semantic memory and social skills — three cognitive domains affected in schizophrenia. (The complexity of humor, however, can also diminish its attractiveness as an experimental cognitive task — simpler cognitive tasks tend to be easier to interpret.) Our research group investigated humor recognition deficits in schizophrenia in order to better understand the condition. I will very briefly review our most recent study (Polimeni et al. 2010).

The centerpiece of our study was our humor recognition test consisting of sixty-four black-and-white single-frame comics from popular artists. Half of the cartoons, chosen randomly, were altered so that the original funny caption was replaced by a non-funny, contextually relevant caption. Participants were not asked to judge whether they found the comic amusing, but whether the comic was in its original form and *intended* to be funny. Notice that this is not a subjective task but a test with definite correct answers. Although the test may not be perfect, its purported validity is based on the fact that almost all healthy people will realize, fairly consistently, when a professional comedian is trying to be funny. (Whether participants themselves find the joke amusing is irrelevant.)

We found diminished humor recognition in the schizophrenia group, a similar result to other studies that have examined humor in schizophrenia using other kinds of tasks (Corcoran et al. 1997; Polimeni and Reiss 2006; Bozikas et al. 2007; Tsoi et al. 2008). During the testing session, we also applied an extensive battery of cognitive tests, each assessing a different aspect of cognitive function.[1] We found statistically significant differences in eight of the eleven tests. In total, sixty people participated in the study: twenty schizophrenia patients, twenty healthy control subjects and twenty psychiatric control subjects. One of our main findings was that deficits in executive function, specifically problems with shifting from one set of ideas to another (Wisconsin Card Sorting Test), most closely correlated with humor-recognition deficits in schizophrenia.

The fact that people with schizophrenia don't always fully appreciate the humorous exchanges in their midst can contribute to feelings of social

---

[1] The tests applied were as follows: Wisconsin Card Sorting Test, Stroop Color-Word Test, Trail Making Test A and B, WAIS-III Digit Symbol-Coding, California Verbal Learning Test-II, WAIS-III Picture Arrangement, and WAIS-III Comprehension.

disconnectedness. Although there have only been a handful of studies investigating the cognitive aspects of humor in schizophrenia, these studies are already beginning to reveal some of the possible reasons for the social impairment associated with the condition.

It is clear that patients with schizophrenia, as a group, do worse on a number of cognitive tasks than healthy people. Because people with schizophrenia also tend to perform worse than psychiatric control subjects, these deficits cannot be explained by medication effects, social circumstances or general emotional distress. However, it is also true that a substantial minority of people with schizophrenia demonstrate no appreciable diminished cognition. In short, neurocognitive studies in schizophrenia certainly hint at a pathological process, but the extent and precise consequences of these relatively modest cognitive deficits are yet to be fully understood.

Odds and Ends: Other Potential Evidence for Pathophysiology in Schizophrenia

Other medical tests have sometimes revealed differences between people with schizophrenia and healthy control subjects. For example, altered electroencephalography (EEG) recordings have been observed in some patients with schizophrenia. An EEG machine measures low-level electrical activity on the scalp originating from neuronal activity just below the skull. Several signature electrophysiological recordings have been detected in a number of schizophrenia studies. For example, reduced mismatch negativity response has been consistently observed in schizophrenia, as well as alterations in P300 event-related potentials (Keshavan et al. 2008). The magnitude of difference is comparable to differences associated with neurocognitive tests (affects sizes of about 1.0 or less). This means there is considerable overlap between schizophrenia and normal groups. Moreover, similarly altered EEG recordings can sometimes be found in other psychiatric disorders, which reduces the specificity of these observations.

Pre-pulse inhibition (PPI) studies in schizophrenia have similarly demonstrated deviations from typical responses. These studies attempt to quantify startle responses in different situations. Under normal circumstances, most people will show a reduction in startle response if a "warning" stimulus precedes the startling experience; this phenomenon is called pre-pulse inhibition. In people with schizophrenia, the reduction in response tends to be less pronounced, especially with auditory stimuli. The responses in relatives of people with schizophrenia tend to fall midway between schizophrenic and normal responses. From a physiological standpoint, it is not known why this occurs in schizophrenia — or even if it is evolutionarily disadvantageous (Keshavan et al. 2008).

Schizophrenia can also be associated with abnormal smooth-pursuit eye movements — the ability to follow the path of a moving object. This deficit has been known about for at least a century (Schmid-Burgk et al. 1982). Again, the real-life ramifications of this ostensible degradation in visual-motor performance are not known.

Finally, obstetrical complications are suspected to have some minor association with schizophrenia (Cannon et al. 2002). For example, poor maternal nutrition could conceivably boost vulnerability to psychosis. Although numerous studies have dedicated themselves to this question, no convincing relationship exists. Across studies, a number of varied results and inconsistencies exist. Whether these supposed positive cases represent true schizophrenia or alternative psychotic variants is an important yet unresolved question.

Genetics of Schizophrenia

Deoxyribonucleic acid (DNA) is a long chain, double-stranded molecule well suited for replicating itself. The human genome contains a total of six billion nucleotide molecules spread over forty-six strands of DNA. For DNA to fully function it requires certain protein affixtures; the resulting combination is called a chromosome. The human genome has two sets of twenty-three chromosomes, for a total of forty-six. Each chromosome has a homologous partner — one originating from your mother and the other from your father. Of the twenty-three chromosomal pairs, one set are the sex chromosomes. Females possess two X chromosomes while males have an X and Y. Classical genetics teaches us that for any chromosomal pair, each maternal and paternal gene may be identical (homozygote pair) or different (heterozygote pair). It is estimated that there are about 30,000 functional genes in the human genome.

In terms of genetics, schizophrenia is shaping up to be more like a complex phenotype than a discernable genetic aberration. Remember that schizophrenia is judged to be over 80 percent heritable — one would expect the heart of the pathology to be evident in the genes. But despite several decades of intensive genetic research, no unequivocal genetic abnormality has ever been found.

The pursuit of a genetic cause of schizophrenia — perhaps a single "schizophrenia gene" — has been underway for several decades. The field of psychiatric genetics is extremely complicated, to say the least. The natural complexity of classical genetics is compounded by the obscurity of how genes and environment interact to ultimately form indistinct behavioral phenotypes. Furthermore, genetic systems cannot be easily exposed in full and therefore are probed using indirect techniques like linkage analysis, association studies, and genome-wide association studies. Superimposed on all this ambiguity are high-powered statistical analyses, which are not always appropriately applied. In the end, very few psychiatrists are well enough versed in all these areas to critically assess the literature. I am not, and acknowledge that I can only recite the latest conclusions from the most highly regarded papers in the field.

A few possible genes have been implicated in schizophrenia — DTNBP1 (dysbindin), NRG1 (neuregulin 1), DAOA (D-amino acid oxidase inhibitor), COMT (catechol-O-methyltransferase), DISC1 (disrupted-in-schizophrenia 1), RGS4 (regulator of G-protein signaling 4), RELN (reelin) and ZNF804A (zinc finger protein 804A). Examining the functions of these genes can perhaps lead us to the underlying molecular pathophysiology of schizophrenia. For example, a variant gene of COMT tends to be present in schizophrenia more often than in healthy controls. This is consistent with COMT's involvement in monoamine (dopamine and norepinephrine) metabolism (Harrison and Weinberger 2005). Recall that dopaminergic changes could be linked to psychotic symptoms.

A number of other candidate schizophrenia genes appear to be involved in neurodevelopmental aspects of brain function. Reelin is believed to regulate neuronal migration and synaptic plasticity. Neuregulin 1 (there are four types, 1 to 4) is involved in synaptogenesis as well as other cellular neurodevelopmental physiology (Craddock et al. 2005). DISC-1 appears to be involved in axonal growth and neural positioning (Sullivan 2005). Although less is known about their precise functions, dysbindin, RGS4 and DAOA each seem to be involved in neural expression.

Some, but not all, of the candidate schizophrenia genes overlap with candidate bipolar disorder genes, suggesting a partial genetic relationship between the two conditions. DISC-1, NRG1, DAOA, COMT and ZNF804A may be susceptibility genes for schizophrenia as well as bipolar disorder (Owen et al. 2007; Craddock et al. 2005; Owen et al. 2009). Kishimoto and others (2008) have argued that dysbindin may specifically confer susceptibility to psychosis through its apparent association with schizophrenia, bipolar psychosis and methamphetamine psychosis in ordinary people.

There has been recent interest in a genetic phenomenon called copy number variation (CNV). Recall that there are usually two copies, one from your mother and one from your father, of any particular gene. It has been recently discovered that perhaps as much as 10 percent of genes deviate from this classical two-copies-per-chromosomal-pair framework. Instead, there could sometimes be one, three or four copies of a specific gene. This phenomenon, copy number variation, may facilitate evolutionary processes by broadening species variation. It could also explain the presence of less-than-ideal phenotypes or even diseases. Some evidence suggests a possible connection between the occurrence of copy number variants and schizophrenia, although so far this association appears weak. Furthermore, *de novo* copy number variants — variants which are not inherited from either parent — do not appear to be strongly associated with schizophrenia (Owen et al. 2009), which means that even if CNVs have something to do with schizophrenia, the genetic transformations would have occurred long ago.

In recent years, it has become increasingly evident that the genetic profile of schizophrenia (and bipolar disorder) better resembles phenotypic variation than genetic abnormality or disease (Crespi et al. 2007). None of the genes tentatively associated with schizophrenia seem to be exceptionally uncommon or sinister. The best guess is that schizophrenia could be caused by a combination of 10–30 genes, each individually having a relatively small effect on the total number of schizophrenia cases. In discussing candidate genes of schizophrenia, Riley and Kendler (2006) suggest "the liability variants in these genes are generally expected to be within the range of normal human variation and to have low risk associated with them individually." Citing a recent schizophrenia genetic study involving the ZNF804A gene, Williams and others (2009) state, "The implications for pathophysiology await characterization of the function of the encoded protein, but more generally, since that study strongly shows that one or more common risk variant exists, and therefore escapes purification by natural selection, it refutes the hypothesis that schizophrenia risk alleles are necessarily rare." In a similar vein, Owen and others (2009) conclude, "in no case have specific risk alleles yet been unambiguously implicated as causal and in no case does the strength and consistency (some alleles or haplotypes across studies) of the genetic evidence equal that for genes now known to be involved in other complex disorders." Thus, the genetic evidence is more in keeping with schizophrenia as a variant of normal than a gross genetic abnormality.

A complicated multi-step genetic study by Khaitovich and others (2008) attempted to explore the possible evolutionary origins of schizophrenia. Probing evolutionary timescales in the order of millions of years, one component of the study compared brain metabolites of chimpanzees and rhesus macaques, as well as humans. The authors concluded, "We find that both genes and metabolites relating to energy metabolism and energy-expensive brain functions are altered in schizophrenia and, at the same time, appear to have changed rapidly during recent human evolution, probably as a result of positive selection." The authors suggest that their results are "consistent with the theory that schizophrenia is a costly byproduct of human brain evolution." However, the results seem equally consistent, if not more so, with the shamanistic theory, as we shall see later.

The Nature of Classical Medical Diseases

Evolutionary principles are crucial to our understanding of medical diseases. Every single medical concern becomes, perforce, inextricably linked to an organism's evolutionary history by virtue of the fact that every facet of an organism is the product of evolution. Some of the pioneering psychiatrists such as Sigmund Freud and Karl Kahlbaum understood this lucidly, as far back as the late nineteenth century.

In 1994, Randolph Nesse and George Williams wrote a book entitled *Why We Get Sick*, an ambitious attempt to understand all diseases using an evolutionary framework. Nesse (2005) has since slightly refined his previous work, placing all diseases under the rubric of six evolutionary explanations. I have played with his model, plugging in various diseases, and cannot find any compelling exceptions; it seems to be a satisfactory model to understand the evolutionary raison d'etre of any medical ailment, including schizophrenia. Nesse's six explanations are: novel environments, competition between organisms, tradeoffs or design compromises, evolutionary constraints, primacy of survival and bodily defenses mistaken as diseases.

First I will describe each component of Nesse's sixfold evolutionary disease model. Then I will explore how a number of common medical conditions, including schizophrenia, fit into this model.

Novel environments

Environments constantly change and consequently, challenge the adaptability of organisms. For example, in my home province of Manitoba, Canada, global warming is threatening a number of species including the polar bear. Similarly, in many human populations the transition from a hunting and gathering lifestyle to Western culture has introduced a number of diseases that were previously rare. Our sedentary lifestyle and easy access to highly caloric foods has, in essence, created a toxic environment. Before the last few thousand years, primates rarely had the opportunity to unceasingly eat as much fatty food as desired. Therefore, our bodies were not designed to deal with these excess calories. Ailments such as atherosclerosis, heart attacks, strokes and type 2 diabetes are associated with excess food consumption and accordingly were rare in traditional societies. For example, obesity is conspicuously absent in the photos of indigenous North Americans before the twentieth century. Hypertension, appendicitis, cholesterol gallstones, diverticulitis, gout and carcinoma of the colon are also diseases primarily associated with industrialized societies (Bickler and DeMaio 2008; Kim et al. 2003b).

Modern Europeans and their descendants demonstrate relative protection against ailments such as diabetes, vascular disease and lactose intolerance, compared to many aboriginal groups (e.g., American, Canadian or Australian) (Brown 2009; Johnson et al. 2009). Although Europeans are still somewhat vulnerable to the toxicities of the modern diet, a few positive genetic changes have already taken place within the last few thousand years, allowing some adaptation to Western life.

Competition between organisms

The danger presented by microbial pathogens or cutthroat competition from an invading species can drastically endanger survival, often requiring radical evolutionary responses. As Nesse puts it, such problems "take resources away from other essential tasks and yield mechanisms likely to be dangerous to host as well as pathogen." Nesse points to our often overly aggressive immune system, which can be accidently unleashed upon itself, leading to autoimmune disorders such as systemic lupus erythematosus, Grave's disease or rheumatoid arthritis.

Other diseases that seem to fit this rubric are sickle cell anemia, thalassemia, cystic fibrosis and Tay-Sachs disease. The genes that cause sickle cell anemia, for example, are also protective against malarial infection (Allison 1954; Ashley-Koch et al. 2000). Another ailment related to the battle against malaria is thalassemia (Flint et al. 1986; Clegg and Weatherall 1999). Sickle cell anemia reflects the mutation affecting hemoglobin among African populations while thalassemia reflects the European mutation, which is slightly different. Cystic fibrosis is a respiratory disease that may owe its existence to genes that confer protection against cholera (Betrenpetit and Calafell 1996; Rodman and Zamudio 1991; Schroeder et al. 1995). Likewise, Tay-Sachs disease is believed to have originated in tuberculosis resistance (Dean et al. 2002; Spyropoulos 1988). Each of these genetic diseases appears to be a less-than-ideal evolutionary adaptation to protect against pernicious infectious diseases. According to molecular anthropologist, Sarah Tishkoff, "It is hard to imagine anything else that can wipe out a population so rapidly. Only those people who have some sort of resistance survive and pass on their genes" (Bradbury 2004).

Certain adaptations to infectious agents possess an obvious commonality – they are terribly imperfect. Ailments such as sickle cell anemia or thalassemia can be seen as desperate stopgap measures to contend with the sudden attack of lethal infectious agents. Because of this evolutionary race against time, any positive genetic mutation, no matter how imperfect, will be superior to no response. I will later argue that certain characteristics of schizophrenia, like paranoia, could reflect less-than-ideal evolutionary adaptations against an equally critical threat: warfare from other hominoids.

Tradeoffs or design compromises

The body of every animal is a survival machine incorporating hundreds of tradeoffs to contend with conflicting evolutionary demands. These tradeoffs can result in disease or vulnerability to injury. For example, muscle strains tend to occur during overwhelming physical activity. However, the size of each muscle group is the result of evolution striking a balance between the advantage of strength and the disadvantage of needing to forage for energy to sustain those muscles. Similarly, our predilection towards knee osteoarthritis or vertebral disc herniation, compared to other ambulatory mammals, is mostly related to the increased stress upon certain joints due to upright posture. Almost every nook and cranny of an animal or plant provides examples of balancing tradeoffs.

Evolutionary constraints

Evolutionary changes are the accumulation of one convenient genetic accident upon another. Evolution does not plan for the future — the principal requirement for any organism is to survive the day. Individuals are therefore prisoners of their evolutionary history and their corresponding genetic make-up. Ripping up the genetic blueprints and starting from scratch is not an option.

The recurrent laryngeal nerve is perhaps one of the best examples of a nonsensical legacy of evolution. Instead of going directly to the larynx from the brain, the laryngeal nerve circuitously loops under the aorta before arriving at the larynx. There is no physiological purpose for this detour; it exists because of anatomical constraints during early embryonic development. In giraffes, the recurrent laryngeal nerve runs an extra five meters!

Another example is the eye's blindspot. The blindspot or optic disc is the location where the optic nerve exits the retina — smack in the middle of our visual field. Again, this is a flaw resulting from the natural constraints of evolutionary history. The point is that almost every disease could conceivably have been avoided with a better design (if only organisms didn't need to evolve, and instead could have been designed by some omnipotent creator).

Primacy of survival

In the world of evolution, survival is king and to ensure survival, evolution must sometimes act drastically. Suffering is one of these desperate measures: our genes pretty much program all of our pain and agony. Suffering improves survival by ensuring that mobile organisms avoid dangerous predicaments. (Plants, by contrast, don't require pain receptors.) Humans, for example, are exquisitely sensitive to pain from burns upon their skin. This is a result of two factors: the commonality of fire, and its extreme destructiveness. If fire were either rare or innocuous, the corresponding pain sensors would not be as jacked up. The most important parts of our anatomy such as the head, hands and scrotum are packed with many more pain sensors than, for example, our lower limbs.

Similarly, in order to prevent a person from bleeding to death after a serious laceration, blood must be able to coagulate very rapidly. However, having a hair-trigger coagulatory system also contributes to such problems as pulmonary embolism or deep vein thrombosis. Clearly, evolution is not for sissies.

Bodily defenses mistaken as diseases

Symptoms such as vomiting or coughing are sometimes confused with genuine medical problems. In fact, they are desirable physiological responses to contend with threatening situations. For example, vomiting is beneficial because it jettisons ingested poisons, while coughing expels upper respiratory irritants.

Almost every medical complaint seems to fit into one of Nesse's six categories. For example, any minor or rare medical condition that does not significantly affect evolutionary systems usually represents a design compromise. However, three well-known medical problems cannot be so easily categorized into Nesse's formulation. Cancer, dementia and certain psychiatric ailments require some supplementary explanations.

Cancer's relationship to evolution is not at all obvious. Cancer rarely affects people before childbearing age — this is in accordance with evolutionary theory: catastrophic design flaws may be tolerated only if they occur infrequently. The drastic rise in cancer rates which begins in late middle age and accelerates as we age could confirm that our survival beyond childbearing years is evolutionarily unimportant. The high frequency of cancer among the elderly may mean that it is related to programmed death. Modernity does not appear to be the primary cause of cancer — animals get cancer too, and traditional societies also faced carcinogenic risks such as persistent smoke exposure.

Animals do not become old because they have become weathered by years of service. Instead, it seems that our bodies are programmed to expire after a certain period of time. This median lifespan is different for every species (Mitteldorf and Pepper 2007; Mitteldorf 2004). In contrast, the most primitive species such as algae replicate indefinitely and therefore possess immortal cell lines. Programmed death cannot be explained by the evolutionary mechanism of individual selection — plain and simple. It is far more likely that some sort of species-, family- or phylum-level selection (group selection) occurred at some juncture in evolutionary history. The fact that no mutation has ever resulted in an immortal animal could suggest that immortality is a devastating problem for the survival of a species, because it essentially shuts the door on variation. It is conceivable that evolution contains several checks and balances to ensure immortality never occurs in certain species, and cancer may be one such mechanism.

It is also not clear how dementia is compatible with evolutionary theory. Is it a design flaw that is tolerated because it occurs during the last stages of life? Or, because human beings are typically living longer, is it a disease of novel environments (Ghika 2008)? These questions await further research.

The following psychiatric conditions also require further examination in terms of Nesse's evolutionary disease formulation: anxiety, depression, bipolar disorder, obsessive-compulsive disorder (OCD), autism and schizophrenia. These ailments are both conspicuous and common. By "conspicuous," I mean that each condition is associated with a set of behaviors so obvious that they would seem to be evolutionarily disadvantageous. By "common," I mean substantially greater than typical mutation rates, which implies that evolution is not effectively discarding the phenotype. These two factors taken together mean the possible effects of evolution cannot be reflexively dismissed. Depression and anxiety could fall under Nesse's category of "primacy for survival": anxiety may deliver its advantages through avoidance of dangerous situations, and Price and others (2004) suggest that depression may cause its sufferers to signal appeasement in social competition, which may lessen aggressive responses from other humans.

The possible evolutionary advantages of the remaining four psychiatric conditions — bipolar disorder, OCD, schizophrenia and autism — are harder to explain away. Each ailment is seemingly heritable, common and moderately conspicuous (mild autism and OCD less so). Moreover, these conditions affect individuals during their prime reproductive years. Of the four ailments, I would argue that schizophrenia is the most conspicuous and socially disruptive. Therefore, from an evolutionary disease perspective, schizophrenia could be unique.

Most of the current evolutionary interpretations of schizophrenia rely on the evolutionary tradeoff clause. The usual argument is that schizophrenia is the evolutionary cost of an advanced brain — specifically a creative brain (Crow 1997; Horrobin 1998; Khaitovich et al. 2008). This is not an unreasonable argument per se; however, when one carefully inspects the pertinent epidemiological parameters, schizophrenia has absolutely no parallel. By contrast, vertebral disc herniation (an evolutionary cost of walking upright) inflicts fewer people, occurs later in life and is generally less debilitating than schizophrenia, thus limiting its adverse effects on reproductive potential. If schizophrenia were strictly a purposeless medical disease, it might be the maximum tradeoff ever tolerated by nature. In other words, in order to invoke the evolutionary tradeoff model, one may have to concede that schizophrenia is its most extreme example. Although this does not disprove the tradeoff argument, it strongly suggests that something evolutionarily positive accompanies schizophrenia.

According to Nesse, the boundary between disease and phenotypic advantage is much murkier in nature than in the physician's office. There is not always a distinct line separating adaptive phenotype from disease, and neither label precludes the other. Organisms scratch and claw for survival and consequently, not every evolutionary solution is elegant. While it is true that nature does not offer countless examples of such disease-phenotype hybrids, several do exist. In addition to sickle cell anemia, thalassemia and cystic fibrosis, another adaptive phenotype-disease hybrid is exemplified by dwarfism in African Pygmies.

In Western populations dwarfism is frequently associated with hypopituitarism — undoubtedly an endocrinological malfunction. However, dwarfism in African Pygmies appears to be an evolutionary adaptation (Shea and Bailey 1996). It has been surmised that diminished stature reduces caloric intake while facilitating the navigation of extremely dense forests. Evidence suggests that African Pygmies constitutionally possess reduced numbers of growth hormone receptors in their bodily tissues (Baumann et al. 1989; Merimee et al. 1989). Evidently these retrograde physiological changes represent an evolutionary adaptation rather than a medical disease. In theory, schizophrenia too may possess certain retrogressive traits that could paradoxically be evolutionarily advantageous.

Neuropsychiatric Conditions

Another way to organize neuropsychiatric diseases involves aligning categories based on their supposed genetic etiology. Using this scheme, schizophrenia again shows itself to be a medical ailment that doesn't run with any other pack of medical diseases. In their chapter "Treating Neurological and Psychiatric Diseases" Cooper and others (2003) organize all neuropsychiatric diseases into three genetic categories: chromosomal abnormalities, single-gene defects and polygenic diseases.

The first category, chromosomal abnormalities, are a type of genetic malfunction and therefore, unambiguously, medical disease. For example, trisomy 21, also known as Down syndrome, is relatively rare, affecting about 1 in 1000. It is associated with mental retardation and other behavioral abnormalities — nothing that could be construed as being beneficial. Fragile X syndrome, another genetic abnormality, can also result in mental retardation. Williams syndrome (also known as Williams-Beuren syndrome) involves a deletion of part of chromosome 7 and is similarly associated with mental retardation. There has been some literature suggesting that Williams syndrome could be associated with enhanced musical skills (Levitin 2006; Dykens et al. 2005; Lenhoff et al. 2001; Don et al. 1999), but under greater scrutiny, these supposedly positive studies tend to be poorly designed. In fact, a double-blind study with fourteen subjects and fourteen controls, clearly the most rigorous of the bunch, revealed similar or slightly diminished musical skills in people with Williams syndrome (Hopyan et al. 2001).

The second group of neuropsychiatric disorders is the single-gene defects. Cooper and others list thirty-two single-gene neuropsychiatric disorders, including Hartnup disease, metachromatic leukodystrophy, cretinism, Hurler's disease and Wilson's disease. Each of these genetic malfunctions is associated with mental retardation or other behavioral abnormalities. No symptoms related to these disorders have any sign of being evolutionarily advantageous. Moreover, they are all rare, which is in accordance with evolutionary theory.

Polygenic diseases — conditions supposedly caused by more than one gene — make up the last category of neuropsychiatric diseases. According to Cooper and others, the most likely polygenic conditions are multiple sclerosis (MS), epilepsy, depression and schizophrenia, although there are probably others such as Alzheimer's disease, bipolar disorder, autism, and obsessive-compulsive disorder.

MS is undoubtedly an enigmatic disease. However, the prevalence and heritability of MS are lower than schizophrenia, making an evolutionary comparison less relevant.

Epilepsy is not a specific medical disease but rather a collection of seizure syndromes which originate with a variety of neuropathological etiologies. The potential causes of epilepsy include head injury, brain tumors and idiopathic origins. Because of the multitude of potential etiologies, some forms of epilepsy may be highly heritable, while others not at all. Epilepsy is also substantially less common than schizophrenia and tends to predominantly affect the very young, as well as the very old (Fong et al. 2008). It would therefore appear that fecundity is less compromised in epilepsy than in schizophrenia.

Organizing neuropsychiatric conditions into groups that reflect their supposed genetic etiologies is a useful exercise because it provides another framework to appreciate the nature of neuropsychiatric disease. Again, schizophrenia, bipolar disorder, OCD and autism look like very unique phenomena, dissimilar to other genuine diseases.

Psychosis Associated With Medical Conditions

Three medical conditions, epilepsy, dementia, and delirium, require special examination because psychosis is sometimes a part of their symptom complex. These medical conditions hinder higher-order cognitive processes and therefore each one is almost certainly a disease or serious biological malfunction. Moreover, there do not appear to be any redeeming qualities associated with any of these ailments — one does not become faster, stronger or mentally sharper when stricken. The fact that psychosis often accompanies these well-known diseases implies that insanity may be a pathological process — a sort of guilt by association.

However, psychotic symptoms secondary to medical disorders may not be exactly the same as those bizarre delusions and hallucinations frequently observed in schizophrenia. Such subtle discrepancies in symptomatology could be representative of veritable underlying differences that separate pathological psychosis from evolutionarily adaptive insanity.[2] To best explore possible differences between psychosis associated with medical diseases and schizophrenia, I have divided the remaining section into three parts: Psychosis in Epilepsy, Psychosis in Delirium and Dementia, and Treatments for Psychosis Based on Disrupting Neural Circuitry.

Psychosis in Epilepsy

Epilepsy is considered a heterogeneous medical phenomenon because it possesses several dozen etiologies. The presentation of epilepsy can range from dramatic whole-body grand mal seizures to subtle absence (petit mal) seizures. Psychosis in epileptic patients has been observed since antiquity (Hyde and Weinberger 1997; Sachdev 1998; Kanner 2000). In a recent study, 5.4 percent of patients from an epilepsy clinic had psychotic symptoms (compared to 0.17 percent in a general medical clinic) (van der Feltz-Cornelis et al. 2008). Although frontal lobe seizures can produce psychosis (van der Feltz-Cornelis et al. 2008), temporal lobe epilepsy appears to have a stronger association with psychosis (Elliot et al 2009; Nadkarni et al. 2007). Treatment with anticonvulsants is usually more effective than

---

[2]People with several rare medical conditions such as velocardiofacial syndrome, Fahr's disease and Friedreich's ataxia seem to display psychotic symptoms more often than the general population (Sachdev and Keshaven 2010). However, the precise nature of the associated psychosis has not yet been well characterized, which places doubt on whether such symptoms can be considered "schizophrenia-like."

conventional antipsychotics. The precise relationship between seizure and psychosis is not known. In other words, it is not known whether seizures directly induce psychotic experiences, or whether brain lesions produce both psychosis and seizures. Although both circumstances seem possible, there is good evidence of the former in at least a subset of cases (Adachi et al. 2002; Umbricht et al. 1995; Roy et al. 2003).

Spiritual experiences have had a special association with epilepsy since ancient times. Contemporary epidemiological studies have confirmed this link (Ogata and Miyakawa 1998; Devinsky and Lai 2008). Hyper-religious experiences are quite commonly associated with seizures, particularly with temporal lobe epilepsy (TLE) (Dewhurst and Beard 2003). It has been noticed that a small subset of epileptic patients, around 1 to 5 percent, experience religious delusions or hallucinations (Ogata and Miyakawa 1998; Devinsky and Lai 2008). Although the percentage is small, these religious conversions are often dramatic and therefore readily noticed. Studying this matter directly, Brewerton (1997) reported on a sample of ten patients deemed to have psychotic symptoms secondary to complex partial seizure disorders. He found that seven of the ten had religious delusions or hallucinations. Upon closer inspection, using a broader definition of religiosity, two of the three remaining patients that were originally classified as not having religious delusions, actually did have spiritual delusions. This could mean that as many as nine of ten episodes of seizure-induced psychosis include religious or spiritual themes. The obvious question is why.

Seizures represent an abnormal or inappropriate activation of neural networks. The uncoordinated electrical activity corrupts the expected outcome of neural processes which support motor, sensory or experiential functions. For example, a localized motor seizure may trigger a muscle group, resulting in a simple twitch which resembles ordinary muscle activity, or it may set off a widespread grand mal seizure. In addition to affecting motor functions, seizures can produce sensory or experiential phenomena: sensations such as anger, fear, déjà vu or religiosity have been associated with TLE.

The presence of a religious delusion or hallucination during a seizure does not mean that psychosis is itself a pathological process. In fact, it could just as easily mean the opposite, by implying the existence of established neural networks which support religious or psychotic experiences. We know that there are dedicated neural circuits in place that produce fear or anger — because ordinary religiosity is so common, some neuroscientists have pondered the existence of assigned neural circuits that support feelings of spirituality. (I will review the neuroscience of religion in Chapter 7.)

Absurd religious thoughts and other possible forms of psychosis are neither exceedingly rare nor are they enormously common. Therefore a fair question is, does psychosis originate in committed neural pathways or is it just an accidental spin-off from other unrelated networks?

Psychosis in Delirium and Dementia

Psychotic symptoms manifest themselves differently in delirium than in schizophrenia. For example, visual hallucinations appear to be more prevalent in delirium. Such disparities can also be seen in the various types of paranoid delusions. Patients with schizophrenia are typically paranoid of nebulous otherworldly entities like devils or aliens, while delirious (and demented) patients tend to make misattributions or have unrealistic suspicions related to ordinary social interactions.

In a retrospective study involving 227 cases of delirium, about 50 percent of cases demonstrated psychotic symptoms. About one-quarter of the entire sample experienced delusions and 62 percent of those delusions were of the paranoid type (Webster and Holroyd 2000). Religious delusions were not mentioned at all (although the authors did not appear to specifically tally religious delusions). These results are in accordance with my own clinical experience: I have noticed that psychosis related to delirium (or dementia) is rarely spiritual in nature. Another interesting observation from the Webster and Holroyd study was that patients with bipolar disorder or schizophrenia did not demonstrate increased rates of delirial psychosis, suggesting that "the pathophysiologic causes of psychosis in dementia, bipolar disorder, or schizophrenia are different than that found in delirium" (Webster and Holdroyd 2000).

Psychotic symptoms are also fairly common in dementia (Bassiony et al. 2000; Ropacki and Jeste 2005); however, spiritual delusions do not seem to be mentioned in the literature. My clinical experience is in complete agreement with the psychiatric textbooks, which state that delusions in dementia are rarely as bizarre as those found in schizophrenia, but instead tend to reflect suspiciousness or "misidentification" of familiar experiences (Lautenschlager and Kurz 2010).

Treatments for Psychosis Based on Disrupting Neural Circuitry

In a fascinating article by Malur and others (2000), the authors identified five case reports of delirium caused by medical illness in patients with schizophrenia or bipolar disorder coinciding with a reduction of psychotic symptoms. They argued that a number of historical treatments for psychosis (i.e., insulin coma treatment, electroconvulsive therapy) produced their therapeutic benefits through inducing delirium. In a similar vein, Bhugra and Potts (1989) reported on two cases where protracted psychosis remitted after significant burn injuries. The clear implication is that people with schizophrenia may need to be healthy in order to be psychotic. Such observations are consistent with my own clinical experience. For example, I cannot recall one instance when a patient with chronic schizophrenia worsened after admission to a medical ward.

Insulin coma treatment began in the 1930s and was used to treat intractable psychotic symptoms in schizophrenia. Inducing coma was dangerous and the treatment was sometimes associated with death — perhaps as high as 1 percent (James 1992). It was abandoned by the late 1950s with the advent of antipsychotic medications, which were undoubtedly safer.

The therapeutic mechanism behind insulin coma treatment is not known, although delirium induction may be the front-running hypothesis. Some have questioned whether insulin coma treatment was ever truly effective (James 1992; Ackner et al. 1957). This question may never be fully resolved because older studies, whether in favor or against, tended to use poor methodologies. Certainly, many front line clinicians believed it was helpful for some patients.

Psychosurgery for schizophrenia is an historical intervention that seemed to reduce certain symptoms of schizophrenia — at a tragic cost for most patients. Tens of thousands of lobotomies were performed in the 1940s and 1950s. One of the early pioneers of psychosurgery, Egas Moniz, framed the problem of psychosis this way: "It is necessary to alter these synaptic adjustments and change the paths chosen by the impulses in their constant passage so as to modify the corresponding ideas and force thoughts along different paths..." (Berrios 1997). In fact, this is precisely the intention when contemporary psychosurgery is used to treat severe refractory obsessive-compulsive disorder (Kim et al. 2003a; Dougherty et al. 2002; Tye et al. 2009). By disrupting the supposedly deviant neurocircuitry underlying obsessive-compulsive disorder, symptoms can be lessened.

Conclusion

This chapter explored the pathophysiological findings related to schizophrenia and sought to fit the condition into an evolutionary framework for general medical diseases. In addition, we compared psychotic symptoms related to schizophrenia to psychosis associated with ordinary medical ailments. Such an examination can help determine whether schizophrenia is truly a disease.

There are a few neuroanatomical and neurophysiological findings which suggest brain dysfunction in a substantial number of schizophrenia cases. For example, diminished size of prefrontal or temporal lobe structures implies neurological malfunction. However, the extent of such deficits may be due to other factors disproportionately associated with schizophrenia, like poor nutrition, depression, illicit substance use or antipsychotic medications. Moreover, it is well known that schizophrenia has no specific pathognomonic findings. In other words, there is no singular pathological mechanism that defines schizophrenia. This is in contrast with most medical diseases, which are almost always associated with specific causal mechanisms.

Although there is mounting evidence pointing towards a polygenic etiology, schizophrenia finds itself apart from other classic polygenic diseases. Many polygenic diseases, like Type 2 diabetes or cardiovascular diseases, are the direct result of toxic modern environments which make them uncommon in traditional societies. The few remaining medical diseases believed to be polygenic are either relatively uncommon, like multiple sclerosis, or do not affect persons in their childbearing years, like Alzheimer's dementia. In accordance with evolutionary theory, one would not expect the long-standing existence of a high-frequency polygenic *disease* accompanied by considerably reduced fecundity.

Psychotic symptoms are observed in such disease states as epilepsy, dementia and delirium. This guilt by association implies a pathological mechanism, but cannot be considered absolute proof that schizophrenia is a disease. By analogy, vomiting resulting from a brain tumor is a pathological symptom, however to puke after drinking an entire bottle of tequila is an evolutionary adaptive response. In fact, there is evidence that schizophrenic psychosis is unlike other types of derailment from reality. Moreover, the way in which schizophrenic psychosis is different — often through fixations with religious ideas — implies an evolutionarily adaptive phenomenon.

– 5 –
The Silver Lining of Psychosis

Clinicians have the understandable tendency to focus on ailments and problems rather than special aptitudes or extraordinary talent. However, the medicalization of troubling conduct can blind us to the positive aspects of certain multifaceted behavioral traits — a perspective that Jay Belsky, a child psychology researcher, discovered when he began thinking about child development in evolutionary terms (Belsky 1997). Belsky began noticing that children depreciatively labeled "vulnerable" (i.e., children who demonstrate high negative emotion in less-than-ideal environments) appeared to benefit disproportionately in especially supportive conditions. Belsky and colleagues reanalyzed data from a number of studies that previously reported only adverse behaviors, to the exclusion of their own positive data (Belsky et al. 2007). Later studies, fully dedicated to this question, appear to support Belsky's differential susceptibility theory, which outlines how highly sensitive children may struggle more than other children in poor environments, but outshine their counterparts under supportive parenting (Ellis et al. 2011; Pluess and Belsky 2009).

Similarly, a number of classic psychiatric disturbances have prospective bright sides to them. For example, milder forms of bipolar spectrum illness and attention deficit disorder have been associated with creativity (Abraham et al. 2006; Carson 2011; Healey and Rucklidge 2006). Both autistic individuals and their relatives sometimes possess exceptional mathematical skills (Baron-Cohen 2006), and obsessive-compulsive traits are often observed in industrious people (Bradshaw and Sheppard 2000; Polimeni et al. 2005).

In contrast, there appears to be no silver lining associated with conventional medical disease, injury or any other form of unequivocal biological malfunction. Medical problems such as heart disease, emphysema, arthritis, diabetes, brain injury or mental retardation syndromes are never associated with symptoms or traits that can be construed as even slightly beneficial. Neither patients nor their relatives accrue any special benefit from such ailments. Even when one imagines such medical problems existing in the prehistoric world of traditional societies, still no potential evolutionary compensation comes to mind.

Because schizophrenia may be connected to a beneficial phenotype like shamanism, it is worthwhile to search for possible advantages associated with it. This chapter will review a number of positive attributes that have been associated with schizophrenia. Such favorable factors could have directly contributed to the qualities inherent in shamanism, but they may have also improved the lives of those genetically associated with the condition (i.e., relatives) and therefore helped maintain the genotype in the community.

The chapter is divided into six sections. First, I will review the evidence demonstrating that psychosis is much more common in the general population than is usually acknowledged. This observation challenges the popular portrayal of psychosis as a rare or alien pathological trait. Section two examines the evidence for increased fecundity in relatives of schizophrenic patients, which may explain the persistence of schizophrenia genes. Section three explores other special qualities in relatives of patients with schizophrenia — how they could be more reproductively successful, possibly through enhanced intelligence or creativity. Section four examines the possible cognitive advantages of mild schizophrenia (schizotypy). Section five explores anecdotal evidence supporting an association between genius and insanity. The last section reviews studies that show enhanced performance in certain highly specific cognitive domains among people with schizophrenia.

Psychosis in the General Population

Psychosis permeates the general population much more deeply than is usually recognized (Stip and Letourneau 2009). Leaving aside the fact that belief in the occult, paranormal activity, eccentric spirituality or even conventional religion may be considered soft psychosis, there is still a substantial portion of the population — perhaps as much as 10 or 20 percent — whose hallucinatory and delusional experiences truly border on classic psychosis. Jim van Os has studied this gray area and proposes that the symptom of psychosis (but not necessarily the disorder of schizophrenia) is a dimensional phenomenon spanning a continuum from normality to flagrant psychosis (van Os 2003; Johns and van Os 2001; van Os et al. 1999). In order to reflect the pertinent literature, the following types of psychotic experience will be dealt with separately: 1) hallucinations, 2) delusions and 3) paranoid delusions (the most common form of psychotic delusion).

Hallucinations

Hallucinations can occur in any sensory domain: auditory, visual, tactile, olfactory or gustatory. In Western societies, the most prevalent form is auditory hallucination. Visual hallucinations may show greater prevalence in certain traditional societies, especially in cultures that revere the inferred messages of nighttime dreams.

Hallucinations secondary to breakdowns in normal physiology have a greater tendency to be randomly expressed in any sensory domain. This seems to be the case in delirium, where hallucinations can occur in the auditory, visual or tactile sensory systems. Alcohol withdrawal and a number of other organic insults often elicit tactile hallucinations, frequently the feeling that insects are crawling all over one's body — a sensation known as formication. In temporal lobe epilepsy, seizures induce a haphazard variety of peculiar sensations and hallucinations. In one study, the aura of eighty-eight temporal lobe patients was meticulously documented resulting in ten reports of taste sensations, ten of olfactory sensations, five of peripheral sensory experiences, ten reports of "hallucinations" (presumably auditory and visual) and reports of a variety of other unusual experiences including thirteen reports of micropsia (a condition where objects are perceived as being smaller than they are) and sixteen reports of déjà vu (Taylor and Lochery 1987). In contrast, schizophrenia and bipolar disorder are chiefly associated with auditory hallucinations — and not just any noise or musical sound, but conversations. In other words, schizophrenic hallucinations speak to us.

Auditory hallucinations may be connected to the universal experience of listening to our own thoughts — that ongoing internal monologue. The amplification of running verbal thoughts, by whatever mechanism, could conceivably be the basis of auditory hallucinations. In fact, auditory hallucinations sometimes occur in those psychiatric disorders characterized by ruminations and self-doubt: major depression, posttraumatic stress disorder, borderline personality disorder and grief reactions. However, grossly psychotic patients usually claim that hallucinations are distinctly "heard" outside their heads, and usually they believe the voices are those of other people — deceased ancestors, people of the opposite gender, non-human entities or even a running dialogue between two other people. In fact, a very recent study reported that only four of fifty schizophrenia patients had difficulty distinguishing their own thoughts from auditory hallucinations (Hoffman et al. 2008). Such reports imply that schizophrenic hallucinations cannot be fully explained as simply the "turning up" of inner thoughts .

Some studies, using exceedingly broad definitions of "hallucination," have reported very high rates in the general population (Posey and Losch

1984; Millham and Easton 1998). For example, a study involving 375 college students tallied 39 percent as having heard their own thoughts spoken aloud (Posey and Losch 1984). This figure seems extreme, and not in accordance with common experience. A number of problems can be recognized in these studies: substance abuse is never satisfactorily taken into account, the common experience of hearing one's name being called in a public place is counted as an auditory hallucination when it should be classified as an illusion, and the possibility of experiencing one's own verbal thoughts is not adequately distinguished from veritable hallucinations (questionnaires cannot easily tease this out). Studies that have attempted to raise the "hallucination threshold" demonstrate rates in the 5–15 percent range, which is in keeping with typical clinical experience (Johns et al. 2002; Tien 1991; Olfson et al. 2002). For example, a study of 1005 adult primary care patients from a Manhattan clinic revealed that 12 percent of patients had experienced at least one distinct auditory hallucination in their lifetimes and 10 percent had experienced at least one visual hallucination (Olfson et al. 2002).

Delusions

The gray area between impeccable logic and delusion is one of the most complex and messy areas in cognitive science. My view is that delusions arise from intense feelings that have escaped the tether of critical thinking. The origin of these feelings and the reasons behind the lack of critical thought vary in each circumstance. But before delving into delusions, it will be helpful to explain what the word *feeling* means.

Although the word *feeling* seems colloquial and "unscientific," this imprecise term may best describe the genesis of most delusions. I define a feeling as the amalgamation of various emotional states, simultaneously modified by previous learned behaviors and the application of logic. A number of fundamental emotions, such as anger, fear, disgust, sadness and happiness exist. Each fundamental emotion presumably has its own dedicated cognitive substructure. Some of these basic emotions can happen concurrently (e.g., anger and fear), and become further modified by previous experience (e.g., *This is the last time he bullies me*) or critical thinking (e.g., *I will count to ten before deciding whether to retaliate*). The sum total of all these cognitive machinations is the feeling.

The story of how humans make complex decisions begins with the behavior of neurologically primitive organisms whose actions are simple stimulus-response functions, essentially determined by reflex. A mosquito, for example, gravitates towards carbon dioxide because it usually means that the blood of the animal exhaling the carbon dioxide is nearby. Insects, for the most part, act reflexively without analyzing situations — that is why mosquitoes can be easily trapped by a device that emits carbon dioxide to simulate the presence of a mammal.

In the course of evolution, a few species have developed bigger brains and greater intelligence. In these "higher" animals, myriad complex stimuli are input and analyzed, so the simple stimulus-response equation has been replaced by a more complicated formula. The fight-or-flight response in mammals, for example, represents an analysis of several sensory inputs (size of the approaching animal, rate of the approach) in addition to previously learned information (*the last time one of these things approached, it gave me peanuts*). In the end, a number of simpler reflexes get integrated into a broader reaction. For humans, feelings represent the conscious awareness of this churning equation.

For humankind, the feeling equation can be modified by experience, personality, moods and chemicals (hormones, medications, alcohol) as well as logic. All feelings are rooted in a limited number of ancient evolutionary concerns — acquiring food, conserving social attachments, enhancing social status, or preserving physical integrity. Every basic feeling represents a path to a feasible solution to an evolutionary problem. For example, dehydration prompts the feeling of thirst, which then directs our attention towards finding something to drink. An insult can threaten our social status and the resulting irritation readies us for potential solutions such as returning the insult. A variety of elemental feelings such as hunger, envy, jealousy, love, tribalism or altruism all had special relevance in traditional societies. I would add anxiety, depression, mania, paranoia, and spiritual (religious) feelings to the list of potentially evolutionarily adaptive emotional states. The feelings associated with these cognitive-emotional states would have been used to solve challenges in our ancestral environment. Some of these other "psychiatric" feelings will be examined in later chapters.

Although feelings point us in approximately the right direction to solve a problem, *Homo sapiens* occasionally applies logic before taking action. Then again, algorithmic calculations can sometimes be so overwhelming that we instead "go with our gut." When feelings are favored over logic, we may, in certain circumstances, call them delusions. The explicit disconnection between simple logic and a cognitive-emotional state (or underlying feeling) can lead us to delusional thinking.

False beliefs can be caused by a combination of factors; for example, ignorance, low intelligence, overly intense feelings, primitive defense mechanisms (denial), superstitious learning or psychosis. Fear of flying, for example, is arguably a form of very mild delusional thinking. Although it may be reflexive to fear heights, most of us dismiss these feelings when boarding a commercial airplane because we recognize the actual statistical risks of flying. But in some people, fear overrides rational action. Some individuals succumb to this fear but acknowledge the inconsistency, while others may distort facts to justify their trepidation. In contrast, few people fret about wearing seatbelts although reason dictates motor vehicle accidents to be the greater hazard. Because highway crashes were never an evolutionary dilemma, humankind has no innate apprehension associated with this modern peril. Seatbelt legislation is arguably required to compensate for our absence of emotion.

Superstitious thinking is another diluted form of delusional ideation. Animal behaviors are partially dictated by the rules of operant conditioning: pressing a bar triggers the delivery of a food pellet, which reinforces the original exploratory (random) behavior. This automatic cognitive algorithm, called operant learning, generates favorable behavior in most circumstances. However, in some situations operant learning produces irrelevant behaviors that defy logic. The gambler who has won a poker tournament wearing a certain pair of underwear may be tempted to wear the same underwear at the start of his next tournament. Although logic dictates that the cards dealt are independent of choice of undergarment, some people cannot resist these irrational feelings.

Primitive defense mechanisms such as denial can also produce gross misconceptions. These false beliefs are not usually classified as delusions because there are no accompanying signs of "craziness" — no disordered thinking, nonconformity or autistic demeanor. I once witnessed an extreme example of denial in a lonely patient who was convinced that I had romantic feelings for her, despite there being only contradicting evidence. She repudiated my disclaimer with the clever twist of logic that as a medical doctor, I was professionally bound to officially disavow any amorous feelings towards patients — so no refutation from me would ever be credible. Although this woman distorted reality to the point of being delusional, she was not "crazy" in the conventional sense of the word.

Spiritual feelings are that subtle sense that a greater, imponderable entity exists somewhere beyond our tangible borders. Spiritual beliefs, likely produced by sporadic transcendental emotional states, have presumably survived into the twenty-first century because contemporary facts and logic cannot completely discount religious premises (since we don't have an inkling about why the universe exists or how consciousness works). These transcendental emotional states seem to be experienced by the majority of people at some point in their lives. In less-scientific cultures, ideas related to magic, the occult, the paranormal, and witchcraft are considered mainstream. (This topic will be examined more closely in Chapter 7.)

The next level of delusionary thinking is traditional psychosis, but with insight. A number of psychiatric patients possess the typical characteristics associated with "craziness" (i.e., disordered thinking, non-conformity, autistic demeanor, bizarre delusions), but also have insight into their struggles with (biologically induced) delusions or hallucinations. Some schizophrenia patients with distinct paranoid ideas, for example, may recognize their fears as being implausible. A combination of reassurance, education about psychiatric illness and antipsychotic medications can all help erode this delusional thinking.

The most severe psychiatric conditions are seen in patients with bizarre delusions or hallucinations without any accompanied insight — for example, the person who casually informs you that he comes from another planet, without any sense that this will be construed as unusual.

In some ways, psychotic thinking resembles the cognitive workings of language. Just as the fundamental ability for language is innate, while the specific words are environmentally determined, the occurrence of delusions is hardwired while the details change from culture to culture. The delusions of ancient peoples never included spaceships, microwaves or spying cameras. Rather, they contained familiar characters like omnipotent jaguars or extraterrestrial eagles.

As we will see, psychosis often follows one of two broad themes: spiritual or paranoid. Moreover, these two psychotic themes are consistently observed across all cultures and epochs.

No one knows precisely how psychotic delusions arise and admittedly every researcher is only groping for answers. My speculation is that some sort of cognitive-emotional substructure exists that actively spews out delusions and hallucinations in a very small minority of people. In addition, there could be an active cognitive process that prevents the application of logic specifically to these delusions. It is as if there are two symbiotic processes in action: a cognitive module that generates certain types of implausible beliefs which originate in distinct psychotic-like feelings, and a purposeful attenuation of logic preferentially applied to psychotic themes.

Faulty reasoning associated with psychosis has been investigated by a number of research groups (see the excellent 1999 review by Garety and Freeman). A frequent finding is that schizophrenia patients, particularly those with delusions, have a tendency to jump to conclusions (Moritz and Woodward 2005; Menon et al. 2006). Schizophrenia may also be associated with a "generalized bias against disconfirmatory evidence" (Moritz and Woodward 2006; Woodward et al. 2008). A few studies have suggested that, specifically during probability tasks, people with schizophrenia tend to jump to conclusions more often than controls do (Garety and Freeman 1999). However, a reinterpretation of some of this data suggested that schizophrenia patients actually made better Bayesian inferences than control subjects because the latter were overly cautious (Maher 1992)! From my perspective, without knowledge of the precise instructions given to subjects, it is impossible to know which experimental group was more accurate. These are interesting studies but obviously we are still at the outskirts of understanding how psychotic patients succumb to implausible thoughts.

On the surface there appears to be a spectrum of false beliefs, from misconceptions to gross psychosis. A number of studies have methodically investigated the edges of this spectrum by looking at delusional beliefs in the general population (Johns and van Os 2001). For example, a large-scale survey revealed that about 50 percent of British adults believed in telepathy and 25 percent in ghosts (Cox and Cowling 1989). Dedicated psychological measures have even been developed for the explicit purpose of measuring delusions in normal populations (Peters et al. 1999). Using the Peters Delusional Inventory (PDI), Verdoux and others (1998) similarly demonstrated high rates of quasi-psychosis in patients in French medical clinics. A number of parameters in the PDI, like worrying about a partner's infidelity or about people looking oddly at you, do not appear to represent explicit psychosis. However, some other questions truly seem to be in the gray areas of psychotic thinking. For example, 47 percent of non-psychiatric patients in these French medical clinics believed in "telepathic communication" and 23 percent believed in witchcraft or the occult.

Paranoid Delusions

Paranoia is a very common subtype of delusionary thinking. The evolutionary roots of paranoia may be traced to extreme vigilance, a ubiquitous cognitive-behavioral stance found in animals that are frequently preyed upon. From an evolutionary perspective, underestimating danger once and being eaten is infinitely more problematic than overestimating danger a hundred times and wasting a little energy by running away. That is why animals like deer, squirrels and small birds are so jittery or, to put it another way, behaviorally paranoid.

A few studies have concluded that subclinical paranoid ideation is common in normal populations and that paranoid ideation may lie on a spectrum rather than represent a discrete pathological category (Ellett et al. 2003; Combs et al. 2002). Some studies have suggested that normal subjects who score high on paranoid scales demonstrate a number of behavioral and cognitive correlates such as making decisions based on less information, making skewed probabilistic determinations, sitting farther away from examiners and taking longer to read consent forms (Combs and Penn 2004). These behaviors probably reflect constitutional factors associated with paranoid stances but could also be the product of negative experiences. One obvious problem with studying the paranoia spectrum in general populations is that of separating misconceptions related to fear (e.g., *my boss wants to fire me*) from overtly bizarre paranoid delusions (e.g., alien abduction). These two types of paranoid delusion could have entirely different origins while superficially appearing to be on a continuous spectrum.

Whether psychosis is only found in core psychiatric disorders (i.e., classic schizophrenia) or at the edge of normal variation may be an important question, because the latter implies that healthy brains are capable of generating psychotic thoughts. Preliminary epidemiological studies suggest that dimensional approaches may be appropriate. Skeptics, however, would not be unreasonable in suggesting that classical psychosis (especially over-the-top psychosis with bizarre themes) is a different animal. In any case, without finely tuned epidemiological studies or a direct knowledge of the inner-workings of psychosis, this will likely remain an open question for some time.

Increased Fecundity in Relatives of Schizophrenia Patients

Although the basic idea of evolution may be simple, the theory represents a distillation of exceedingly complex interactions. We do not yet fully understand all the multi-layered influences that maintain or eliminate genes in nature. Therefore, we cannot be confident that we have a complete list of all the possible evolutionary-genetic mechanisms that would explain the persistence of a phenomenon such as schizophrenia.

One possible mechanism, which we will explore in greater depth in chapter six, is the balanced polymorphism argument: increased fecundity in relatives of patients with schizophrenia may explain the continued existence of schizophrenia genes. There is some evidence for increased fertility in relatives of patients with schizophrenia — although admittedly this evidence is weak. For example, one study using records from rural Ireland found increased fecundity among the parents of identified male schizophrenia patients (Waddington and Youssef 1996). Another small study involving one hundred patients from India found increased fecundity in parents and siblings of patients with schizophrenia (Srinivasan and Padmavati 1997). Yet another study, of Swedish births before World War II, showed a trend towards enhanced fecundity among the offspring of schizophrenia patients (MacCabe et al. 2009).

In contrast, an approximately equal number of studies have *not* shown differences in fecundity (Vogel 1979; Buck et al. 1975; Svensson et al. 2007; Haukka et al. 2003). The reconciliation of these disparate results is not immediately obvious. Most of the positive studies covered rural communities, while the negative studies represent urban populations. However, the greatest confounding variable could be the paradoxical declines in fertility rates among richer nations over the last few generations. Larger families are no longer a direct manifestation of prosperity as they were throughout most of humankind's evolution.

Other Special Qualities in Relatives

Modern fecundity appears to no longer be an accurate reflection of a person's genuine ability for reproductive success. Therefore, other proxy measures may be more appropriate. Jon Karlsson has shown that academic giftedness is preferentially associated with relatives of patients with schizophrenia. Karlsson's studies have been elegant, in part due to their simplicity. One primary advantage has been the use of comprehensive education and medical demographic records from the entirety of Iceland — a culturally homogeneous and stable population of about 300,000 with a high literacy rate. Karlsson simply tallied entries in Iceland's *Who is Who* index and found, "the likelihood of being listed is almost twice as high for close relatives of psychotic patients as it is for the population at large" and "relatives of manic depressives appear to be particularly likely to be included" (Karlsson 1974). In addition, Karlsson's data showed that both patients with schizophrenia and their relatives had about double the chance of being a high school honor graduate than the general population. Furthermore, both persons with schizophrenia and their relatives had more doctorate degrees per capita (Karlsson 1984).

Follow-up studies by Karlsson continued to show a propensity towards giftedness in relatives of psychotic patients. Using the same Icelandic records, relatives of psychotic patients were more than twice as likely to have been authors of at least two books — a statistically significant difference ($p < .01$) (Karlsson 2001). Similarly, Karlsson found that students who were ranked among the top for mathematical skills had an approximately three times greater chance of being admitted for a psychotic condition (Karlsson 1999). This particular study, however, did not separate manic-depressive psychosis from schizophrenia. Along the same lines, the "superkids" study by Kauffman et al. (1979) showed that children of psychotic mothers were often very gifted. Out of fifty-two children (twenty-two control mothers, eighteen mothers with schizophrenia and twelve bipolar mothers), the six highest ratings for social and intellectual function went to children of psychotic mothers. Similar studies with bipolar disorder patients have shown greater creativity in relatives compared to the general population (Richards et al. 1988; Simeonova et al. 2005).

Special Qualities Related to Schizotypy

Schizotype (schizotypy, schizotypal) is a word, derived from "schizophrenic phenotype," typically used to describe subthreshold schizophrenia. Put another way, schizotypal people are seen as having a touch of schizophrenia. Schizotypal symptoms are not those of gross insanity but are nearer to normality on the psychosis spectrum. The following are examples of schizotypal symptoms: a propensity towards unusual sensations or experiences, eccentric tendencies, inappropriate affect, odd beliefs and metaphorical thinking. DSM-IV reserves the diagnostic category schizotypal personality disorder only for persons who find themselves struggling with such symptoms — the diagnosis is not made if the person is not distressed by the symptoms or suffers no appreciable adverse consequences. It is widely accepted that relatives of patients with schizophrenia demonstrate higher rates of schizotypal personality disorder than the general population (Kendler et al. 1981; Baron et al. 1983).

There has been a lot of research exploring whether creativity is preferentially associated with schizotypy. A common observation is that certain creative professions appear to be overrepresented by schizotypal personalities; hence the numerous studies examining this association. If a positive association existed between schizotypal personality and creativity, it could explain the persistence of the schizophrenia genotype.

About a dozen studies have supported a connection between creativity and schizotypy (O'Reilly et al. 2001; Burch et al. 2006; Schuldberg et al. 1988; Krysanki and Ferraro 2007). On the other hand, one recent study (Miller and Tal 2007) questioned this association. In most studies, schizotypy was measured using the psychoticism subscale of the Eysenck Personality Questionnaire. Creativity was often defined by dedicated participation in a creative endeavor such as music or the visual arts, but in some studies creativity was measured using psychological tests such as the Alternate Uses Test, where blind raters evaluate the uniqueness of various responses generated by study participants. The clear trend is that measures of creativity tend to correlate with schizotypy. For example, Schuldberg and others (1988) tested about a hundred college students and found a statistically significant correlation between schizotypy and supposed measures of creativity. Burch and others (2006) also found higher schizotypy scores among a group of about fifty university-educated visual artists compared to fifty non-artistic university-educated control subjects.

Because both creativity and schizotypy are fairly broad terms, newer studies have tried to home in on the precise cognitive factors connecting these two phenomena. For example, Schuldberg (2000) found a special correlation between creativity and positive symptoms of schizotypy, impulsivity and hypomania. Nettle (2006a) similarly found that creative people share an affinity for unusual experiences, in addition to an absence of avolition and anhedonia — states that are arguably the converse of hypomania. Nettle additionally found that while divergent thinking (thinking outside the box) tends to be characteristic of poets and visual artists, convergent thinking (a propensity towards systematizing data) is observed in mathematicians. Mathematicians demonstrate a tendency towards autistic tendencies (lower interest in unusual experiences, less impulsivity, greater conformity, and higher scores on negative symptoms of schizotypy) rather than the expansive personality traits which are often seen in artistic people. The presence of such antipodal differences between two distinct forms of creativity highlights a crucial problem with this sort of research: when contrasting types of creativity are merged into one sample group, it potentially creates cancelling-out effects that obfuscate results. It will obviously require many more studies before we have a better understanding of both creativity and schizotypy, which will then allow these respective cognitive functions to be compared meaningfully.

Daniel Nettle and Helen Clegg (2006) have taken these kinds of studies a step further by adding appraisals of prospective mating success to their measures of schizotypy and creativity. In one study they recruited several hundred participants ranging from ordinary non-artisans to professional poets and visual artists. Mating success was measured in a flexible manner, by either looking at steadiness of relationships or total number of partners. Creativity turned out to be positively correlated with mating success. Certain aspects of schizotypy (i.e., voicing unusual experiences and impulsive non-conformity) also correlated with mating success. Notably, these positive correlations were observed in both genders. Invoking Geoffrey Miller's theory (Miller 2001) that creativity may be sexually selected, Nettle and Clegg proposed, "mate choice is linked to creativity, and creativity in turn to schizotypy and thence to schizophrenia."[1]

Genius and Insanity

The association between madness and creative genius has been surmised since antiquity. Aristotle said, "No great genius has ever existed without some touch of madness." Plato claimed that a poet's inspiration arose during moments of "divine madness" (Ludwig 1992). The Roman philosopher Seneca said, "there has never been a great mind without some degree

---

[1] I am skeptical about the need to frame such results into a sexual selection framework. The science of sexual selection (how we judge mates and for what evolutionary purpose) is complex. My own view is that sexual selection arguments are often forced — favored only because they avoid discourse leading to multi-level selection (a male-female union represents a dyadic group).

of madness" (Bradshaw and Sheppard 2000). But what is implied by this association, and which way does the arrow of causality point?

There are two possibilities that are not mutually exclusive: the cognitive fluidity of madness facilitates creativity, or geniuses are more prone to suffer from madness because they possess hyperactive thoughts and emotions. Skeptics will propose a third possibility: that any connection between genius and insanity is entirely coincidental, and anecdotal accounts are disproportionately recognized because they are so striking and memorable.

Creativity is a colloquial expression broadly applied to a multitude of undertakings. Scientists, inventors, artists, musicians, storytellers, comedians and even some athletes have required creativity for their greatest successes. In my view, divergent thinking is the embodiment of creativity. Novel insights require being "off" from conventional thoughts — but not by too much, or thinking becomes disorganized and ineffective. It seems intuitive that the mental paroxysms of insanity, toned down, could result in a cognitive fluidity suitable for creativity. Creative problem solving seems to rely on a fine balance between divergent thinking and organization. Therefore, we would not expect floridly psychotic people to express creativity effectively. Moreover, the ideal balance may be different for each creative endeavor; visual art may rely on more divergent thinking while mathematics perhaps favors systematized thinking (Baron-Cohen et al. 1998).

There has always been a small segment of society described as odd, off, peculiar, flaky, quirky, different, eccentric, weird. For centuries it has been anecdotally observed that such people are also creative. This oddness is typically not its own separate entity, but reflects a mild form of psychotic spectrum illness, usually bipolar disorder or schizophrenia. Mild psychiatric conditions can include certain symptoms such as hypomania or nonconformity which seem to be especially useful for creative endeavors. Hypomania, for example, boosts energy while enhancing affective attunement — a mental state particularly useful in creative writing and music composition.

At present we simply do not know which neural pathways support creativity. A reasonable hunch is that creativity is related to semantic memory systems. Semantic memory refers to that aspect of long-term memory that integrates ideas, word meanings and concepts. Semantic memory is perhaps best described using connectionist models; that hypothetical cognitive network of interrelated memories briefly reviewed in chapter four (Saumier and Chertkow 2002; Nestor et al. 1998). Recall, in the connectionist model, words, ideas or memories reside in nodal points connected to each other according to the strength of their semantic association.

It is possible that these hypothetical semantic maps represent actual neuroanatomical substructures. For example, upon hearing the term "ocean," the word "blue" will frequently come to mind. It is well known that word association tasks can be made easier by exposure to semantically related words (Minzenberg et al. 2002; Spitzer 1997). Introducing the word "sky" a few seconds before the word "ocean" will often prime the response word "blue," so that it comes to mind several microseconds faster (Nestor et al. 2006). It is as if the priming word (sky) is somehow able to neuroelectrically activate the neuronal environment corresponding to the other words (ocean, blue).

It has been known since the early twentieth century that during word association tasks psychotic patients generate fewer conventional words and more deviant words than normal control subjects (Kent and Rosanoff 1910; Spitzer 1992; Johnson and Shean 1993). Accordingly, schizophrenia patients, especially those with high levels of thought disorder, often fail to be primed by related words (Aloia et al. 1998). Without a "normal" semantic priming system, conventional responses are more likely to be supplanted by secondary, indirect or even unrelated associations (Weisbrod et al. 1998). However, an equal number of studies have shown that people with schizophrenia are sometimes faster than control subjects during certain semantic priming tasks (Doughty and Done 2009; Manschreck et al. 1988). Semantic memory experiments examine extremely complex cognitive processes and as of yet, there is no satisfactory explanation for these discrepant results. Nonetheless, the deviance in semantic memory characteristic to schizophrenia, whatever its precise nature, may be the fountainhead of human originality.

But how does one study the possible connection between insanity and creativity in a methodical manner? This task is formidable due to several vexing obstacles. First, how do you quantify creativity and its relative contribution towards any stroke of genius? This seems to be particularly difficult in the field of science. For example, did Alexander Fleming's discovery of penicillin require creativity or was it a serendipitous observation obvious to anyone well versed in the study of bacteriology? Carl Linnaeus certainly needed organizational skills to establish a taxonomical system for the entirety of nature's organisms, but how much creativity was necessary for the endeavor? On the other side of the spectrum, August Kekulé claimed to have envisaged the chemical arrangement of a benzene ring during a daydream. We have no way to reliably measure the creative portion of any ingenious accomplishment.

Incomplete biographical information about prominent creative people is a second significant hindrance to the study of creativity and psychiatric disorders. It is not uncommon for patients to conceal their most aberrant thoughts or behaviors from everyone except confidants. Moreover, publicized biographical information will often become either romanticized or normalized. Even with the numerous anecdotal reports of psychotic episodes among successful people, we could just be witnessing the tip of the iceberg.

The third obstacle relates to the question, which psychiatric problems are most relevant to creativity? For example, it appears that bipolar II disorders are overrepresented in music and writing, and mathematics may be associated with autistic or schizotypal tendencies. If a researcher compares every form of creativity to every possible psychiatric disorder (including anxiety disorders which may have no bearing on creativity), any possible positive correlation will be diluted. Before a final adjudication can be made about creativity and mental illness, we will have to achieve a more nuanced understanding of how different psychiatric conditions interrelate with the various forms of creativity.

It is also possible that some psychotic episodes attributed to creative people may have been due to chronic alcohol use, poisoning or concomitant medical problems, rather than psychiatric etiologies. General paresis of the insane, one of the manifestations of tertiary syphilis, was sometimes accompanied by psychotic symptoms. On the other hand, some historians have been premature to attribute organic causes to psychotic episodes (Keynes 1980). For example, the old expression "mad as a hatter" overstates the tendency for mercury poisoning to cause psychosis. It turns out that hallucinations and fixed delusions are actually rare (Maghazaji 1974; O'Carroll et al. 1995; Waldron 1983). Instead, mercury poisoning typically causes an organic dementia resulting in disinhibited anger and other irregular behaviors.

Over the last forty years or so, a number of studies have attempted to methodically study the possible link between creativity and mental illness (Andreasen and Powers 1975; Ludwig 1992; Schuldberg 1997; Jamison 1989; O'Reilly et al. 2001; Fisher et al. 2004; Barrantes-Vidal 2004; Preti and Miotto 1997; Richards 1981; Rushton 1990). In general, some correlation is demonstrated but not usually to the extent that one can be confident that an association exists. For example, Post (1994) examined the biographies of over three hundred world-famous artists, writers, scientists and politicians and found functional psychosis in 1.7 percent, and no schizophrenia.[2] Post concludes this represents a low association between creativity and psychopathology. However, I believe the most severe forms of psychosis, such as schizophrenia, would not exist in a group that demonstrates such consistent success throughout their lives.

Two recent reviews have analyzed the literature pertaining to creativity and mental illness (Waddell 1998; Lauronen et al. 2004). Although both

---

[2] Whether politicians should be counted in a creative group is debatable.

studies show much agreement, Lauronen and others conclude that an association probably exists, while Waddell is less convinced. Waddell, however, is too dismissive of certain positive studies. For example, although Karlsson's Icelandic study (1974) has relatively few confounds, it is discarded solely because of its retrospective design. Waddell acknowledges that she may be inclined to question the role of creativity in mental illness because of her concerns that this association may promote "romantic associations" and trivialize mental illness — a concern I do not share. Both reviews are correct to emphasize that substantial methodological flaws exist in most of the relevant studies.

I will now review a number of anecdotal cases of successful people with serious psychiatric ailments. In no way do I suggest that using anecdotes reflects a rigorous examination of the field — the main purpose of these biographical sketches is to raise a few eyebrows. I have incorporated bipolar disorder because of its possible connection to schizophrenia and shamanism. This subsection is divided into the following areas: sports and games, music, visual arts, writing, science and leadership.

Of all the creative outlets, athletics seems to show the least substantive association between insanity and genius. Most of the greatest athletes, like Tiger Woods, Wayne Gretzky, Michael Jordan and Diego Maradona, do not appear to have had major mental illness, and whether creativity was a necessary part of their talents can also be debated. The "Fosbury Flop" may be a rare example of an imaginative sporting brainchild. In 1968, Dick Fosbury revolutionized the high jump by jumping over the bar backwards at the Mexico City Olympics and winning the gold medal. There was, however, no evidence of mental illness in this track athlete.

Pete Maravich, clearly one of the most ingenious basketball players of all time, seemed to have struggled with mild psychiatric symptoms. Magic Johnson, himself one of the most creative passers in NBA history, apparently said, "Maravich was unbelievable. I think he was ... ahead of his time in the things he did." Maravich played at the most unlikely place for basketball, Louisiana State University, but became the greatest scorer in Division 1 U.S. college history. He scored 44.2 points per game — if there had been a 3-point line the number would have been an astounding 52 points per game, or 57 by other accounts. However, it was Pistol Pete's imaginative passing that was particularly eye-catching. He passed in such inventive ways that basketballs would be routinely seen bouncing off unsuspecting teammates.

Maravich's mother undoubtedly had either bipolar or schizoaffective disorder and, tragically, she committed suicide. Maravich was a moody person with frequent highs and lows, and a number of people around him suspected he also had bipolar disorder. At age 35, Maravich heard the voice of God: "It was about 5:50 in the morning, and I heard God speak to me..." (Berger 1999). Maravich had previously dabbled in Eastern religions but became an exceptionally devout Christian after that experience.

Another interesting story surrounds Canadian golfer Moe Norman, an odd and eccentric man who was very likely autistic. He opted to work in bowling alleys for minimum wage during winter instead of practice in southern climates as top golfers normally do. Norman won a number of major Canadian tournaments from the 1950s to the 1980s. Tiger Woods and Lee Trevino have both pointed out that Norman had one of the most consistent swings in golf history.

Bobby Fischer, one of the greatest chess players of all time, almost certainly had schizophrenia. In addition to being a socially awkward loner, Fischer practiced unconventional religions, expressed bizarre political views and repeatedly made unrealistic demands of people around him. Fischer articulated a number of paranoid delusions including that Gary Kasparov, a former chess world champion, was a KGB agent, and that Bill Clinton was a secret Jew who had personally plotted against him (Chun 2002). It appears that Fischer was never psychiatrically treated because his symptoms were minimized and framed as the eccentricities of genius.

Music can be characterized as a form of emotional communication, and accordingly bipolar disorder seems to show greater association with musical talent than schizophrenia does. It has been suggested that Mozart had bipolar spectrum illness, but if he did there is no evidence of psychosis (Huguelet and Perroud 2005). Beethoven struggled with moodiness, irritability and even suicidal thoughts, interspersed with periods of grandiosity and outrageous gregariousness. He expressed both paranoid and religious delusions. His often-disheveled appearance even led to an arrest for vagrancy on one occasion (Hershman and Lieb 1998; Bower 1989).

A number of twentieth century musicians including Kurt Cobain, Phil Spector, Ozzy Osbourne and Axl Rose have been suspected of having bipolar disorder, according to various news reports. Brian Wilson of The Beach Boys probably has schizoaffective disorder, bipolar subtype. It has been reported that Wilson was tormented by auditory hallucinations for decades (Friedman 2009). Most critics point to *Pet Sounds* as Wilson's most inspired album. To understand the extent of Wilson's creativity, consider that Beatles producer George Martin said, "Without *Pet Sounds*, *Sgt. Pepper* wouldn't have happened. *Pepper* was an attempt to equal *Pet Sounds*" (Wilson 2008).

Glenn Gould, a famous Canadian classical pianist, was well known for his odd behaviors and eccentricities; he likely had autism. Syd Barrett, an original member of Pink Floyd, reportedly struggled with schizophrenia (Rolling Stone 2006).

Some of the most innovative painters have struggled with classic symptoms of major mental illness. Vincent van Gogh suffered through at least a dozen psychotic episodes (Hershman and Lieb 1998). He experienced classic paranoid delusions, such as the irrational belief that the police were after him. He would obsessively translate the Bible in idiosyncratic ways. He tended to be moody, irritable, disheveled, odd, standoffish and solitary. His most notable bizarre action was slashing off his ear and presenting it to a local prostitute. Incidentally, two of Vincent van Gogh's siblings also experienced psychotic episodes.

Edvard Munch, well known for his haunting painting *The Scream*, was an irritable and solitary figure. He frequently behaved peculiarly and unpredictably. In midlife, Munch was hospitalized for anxiety and persecutory delusions (Steinberg and Weiss 1954).

Throughout history, creative writers have not-uncommonly been associated with mental problems. The seventeenth-century English dramatist Nathaniel Lee spent several years at Bethlem Royal Hospital (Porter 2002). Prior to his incarceration he demurred, "They said I was mad; and I said they were mad; and damn them, they outvoted me." Kay Jamison, a psychiatrist who has written about her own bipolar disorder, has suggested Alfred Lord Tennyson was probably bipolar II (Jamison 1995; Jamison 1989). Charles Dickens also seems to have demonstrated classic bipolar II symptoms (Hershman and Lieb 1998). Ernest Hemingway harbored bipolar symptoms and committed suicide, as did his father and two siblings. William Blake had visions of God during his psychotic episodes (Lange-Eichbaum 1932, Smith 2007) and Virginia Woolf heard voices in the context of severe depressions. James Joyce was apparently "aloof and cold" and had a daughter with schizophrenia (Andreasen 2000). In later years, he apparently became "increasingly disorganized artistically" (Andreasen 2000). The poets Ezra Pound (Karlsson 1974) and John Milton (Smith 2007) also appear to have experienced psychotic symptoms.

Some of the most original scientific minds have either had major mental illness or it has been evident in their very close relatives. Socrates regularly experienced auditory hallucinations (Smith 2007). Pythagoras had a lot of unusual beliefs; for example, he claimed to have inhabited the bodies of important people from past generations. His mathematical teachings resembled a religious cult — he frequently ascribed mystical properties to various numbers (Van Doren 1991). In addition to being secretive, peculiar and solitary, Isaac Newton experienced persecutory delusions for over a year around age fifty-one (Jeste et al. 2000). He was also abnormally obsessed with religion, to the point of inventing his own absurd version of a biblical prophecy (Hershman and Lieb 1998). Another brilliant physicist, Albert Einstein, had a son with schizophrenia.

The Nobel Prize-winning mathematician John Nash, diagnosed with schizophrenia, was the subject of the 2001 Hollywood film *A Beautiful Mind*. Another mathematician (and philosopher), Bertrand Russell, had a son, granddaughter, aunt and uncle with schizophrenia (Andreasen 2000). Michael Faraday experienced a psychotic episode in his middle years (Karlsson 1974). The Swedish philosopher and scientist Emanuel Swedenborg had religious visions and delusions. (Lange-Eichbaum 1932). Friedrich Nietzsche may have had bipolar disorder, although an organic etiology was equally possible. In any case, he was admitted to a psychiatric facility for paranoid delusions. Gregor Mendel and Charles Darwin each struggled with depression (Karlsson 1974).

A number of political leaders including Caligula, King Ludwig II of Bavaria and King George III seem to have been insane. King Charles VI of France (1368–1422) and his grandson King Henry VI of England (1400s) both seem to have had schizophrenia (Bark 2002). King Charles VI suffered numerous bouts of disorganized thinking and paranoid delusions. He maniacally killed four of his own men during a psychotic episode (Pfau 2008, 62). He also suffered from the outlandish "glass delusion," believing his body would shatter if hit with enough force.

Daniel Schreber, a prominent German judge during the late 1800s, is particularly well known because his case was later analyzed by Sigmund Freud (McGlashan 2009; Smith 2007). Schreber was admitted to asylums on three occasions, suffering with severe symptoms including multiple suicide attempts and catatonic stupors. His memoirs eloquently describe his religious hallucinations and persecutory delusions.

Enhanced Performance in Schizophrenia

A number of fascinating experiments have found that, compared to normal control subjects, patients with schizophrenia sometimes demonstrate superior performance in a few specialized perceptual skills (Dakin et al. 2005; Ebner et al. 1971; Place and Gilmore 1980; Prentice et al. 2005; Wells and Leventhal 1984; Uhlhaas et al. 2006; Cromwell and Held 1969). To my knowledge, similar findings have never been claimed for mental conditions that are clearly pathological (i.e., mental retardation, dementia or head injury).

Several decades ago an interesting study attempted to quantify creativity and test whether patients with schizophrenia were more imaginative than their control counterparts. Using Guilford's Alternate Uses Test, Keefe and Magora (1980) tested four experimental groups: non-paranoid schizophrenia patients, paranoid schizophrenia patients, psychiatric controls and healthy controls. The Alternate Uses Test asks participants to come up with alternative uses for various items. For example, the stimulus "pair of shoes" generated the following respectable responses from some of the schizophrenia patients: "a place to develop mold cultures," "a plant holder" and "a house for mice or gerbils." Blind raters judged the originality of the responses and not unexpectedly, patients with non-paranoid schizophrenia outperformed the three other groups.

Another study, which reviewed high school records, discovered that schizophrenic patients had higher marks in art, language and religion classes compared to their healthy counterparts (Helling et al. 2003).

In a paper entitled "Are people with schizophrenia more logical than healthy volunteers?" Owen et al. (2007) evaluated whether people with schizophrenia possessed superior theoretical rationality. Study participants were presented with two types of syllogisms. One set of syllogisms was illogical but the conclusions were consistent with common sense. For example:

> If the sun rises, then the sun is in the east.
> The sun is in the east.
> Therefore, the sun rises.

Syllogisms in the second group were intrinsically logical but conflicted with ordinary experience. For example:

> All buildings speak loudly.
> A hospital does not speak loudly.
> Therefore, a hospital is not a building.

Although schizophrenia patients were IQ-, gender- and education-matched to the control group, they convincingly outperformed the control group with an average score of 8.76 correct out of 15 versus 6.21 correct. Because scientific discoveries are sometimes inherently logical while simultaneously contradicting convention, this study implicates the non-conforming disposition of schizophrenic patients as a potential contributor to ingenious creativity.

A number of visual-perceptual experiments have implicated subtle neuroprocessing differences that might favor schizophrenia. The Place-Gilmore experiment was particularly interesting because patients with schizophrenia outperformed control subjects by a wide margin during a very specialized visual-perceptual task (Place and Gilmore 1980). In the original Place-Gilmore experiment, participants were asked to estimate the number of lines in various configurations. Each cluster of lines was visible for fifteen to twenty milliseconds (about one-fiftieth of a second), so participants did not have enough time to actually count the number of lines but instead were forced to estimate. Each display contained three, four, five or six lines, presented either in a relatively symmetrical arrangement (Figure 5.1) or in a disarrayed manner (Figure 5.2). Control subjects performed better when the lines were presented in an orderly arrangement, but when lines were displayed in a disorganized fashion, patients with schizophrenia performed better.

Figure 5.1: Two examples of orderly lines from the Place-Gilmore experiment (Place and Gilmore 1980).

The differences in performance reported by Place and Gilmore were astonishing. For example, when judging four disorganized lines, patients with schizophrenia scored 47 percent correct while controls only totaled 18 percent (n=10 patients with schizophrenia and n=10 control subjects). Judging five disorganized lines yielded 30 percent correct for the schizophrenia group and 10 percent for controls. Six lines resulted in 25 percent correct in the schizophrenia group and 0 percent for controls. A few years later the Place-Gilmore experiment was replicated, albeit with smaller differences in the superiority of performance (Wells and Leventhal 1984).

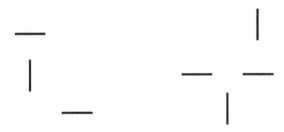

Figure 5.2: Two examples of disorganized lines from the Place-Gilmore experiment (Place and Gilmore 1980).

Uhlhaas and others (2006) used a comparable visual-perceptual paradigm and also showed superior performance in schizophrenia. Their experiment was based on the Ebbinghaus illusion (also known as Titchener circles) (Figure 5.3). This is the familiar optical illusion where a medium-sized circle amidst larger circles appears smaller than the same medium-sized circle among smaller circles. Interestingly, patients with schizophrenia are less fooled by this illusion.

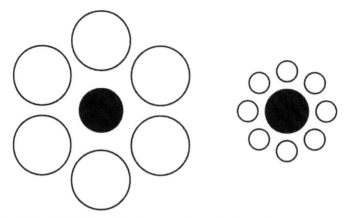

Figure 5.3: The Ebbinghaus illusion; people with schizophrenia seem to be less deceived by this optical illusion.

Dakin and others (2005) used a similar optical illusion (Figure 5.4). In their experiment they asked participants to judge differences in contrast between various images. Again, patients with schizophrenia were less deceived by the distorting effects of adjacent images; in other words, they were much less vulnerable to the optical illusion. Remarkably, twelve out of the fifteen schizophrenics did better than the top-rated control subject.

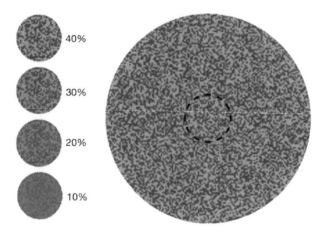

Figure 5.4: Dakin's contrast illusion. The dashed target is identical to the 40% contrast circle. Patients with schizophrenia seem to be less deceived by this optical illusion.

All of these visual-perceptual papers have interpreted their findings using a disease framework of schizophrenia. Thus, the universal assumption has been that some cognitive-perceptual mechanism is broken in schizophrenia. One reasonable interpretation has been that, in certain circumstances, people with schizophrenia inefficiently process contextual information while perceiving complex visual stimuli (Uhlhaas and Mishara 2007; Uhlhaas and Silverstein 2005). However, an equally valid interpretation is that control subjects are more reliant on contextual information and inappropriately invoke contextual information in certain circumstances.

Intrigued by the Place-Gilmore findings, I wanted to replicate the study using a modern neuropsychological software program called E-Prime, and enroll a larger set of participants. Along with colleagues Jeff Reiss, Darren Campbell and Breanna Sawatsky, we tested twenty-seven schizophrenia patients and twenty-nine control subjects. However, we could not confirm the original Place-Gilmore findings. Instead, in our study controls slightly outperformed patients with schizophrenia across the board. I have no explanation for this discrepancy. The reason may be as simple as an unknown computer glitch on our part. Alternatively, the problem may reside with the original Place-Gilmore study and their selection of only ten study participants per group. In such small studies, a few extraordinary participants can drastically skew results.[3]

Conclusion

I have hoped to convince the reader that psychotic thoughts are not a malignant imposition upon the psychology of humankind. In contrast to almost every other genuine medical disease, schizophrenia appears to be associated with a few phenotypic and genotypic benefits.

First, experiences along the psychotic spectrum are a lot more common in the general population than usually recognized. This puts unconditional disease models of insanity in a precarious position because widespread phenotypic traits are almost always evolutionarily adaptive in some way.

Second, there is some evidence that relatives of people with schizophrenia are more creative than average. This quality can also justify the persistence of the schizophrenia genotype.

Last, people with schizophrenia seem to enjoy a few talents of their own. Any special aptitude associated with schizophrenia could be coincidental and inconsequential, or the specialized cognitive endowments

---

[3] Because of the negative result, we did not attempt to publish these findings. (It is, however, acknowledged that such negative results are probably worth publishing.)

possibly associated with schizophrenia may have served some greater evolutionary function.

Before we delve into the heart of the shamanism question, a few more topics require attention. In the next chapter, the basic mechanisms of evolution will be examined, as well as the primeval circumstances that shaped modern humans.

– 6 –
All Things Evolution

The forces of evolution have been shaping animal brains for a very long time. Some of the fundamental units of the mammalian nervous system first appeared several hundred million years ago. Take, for example, the nicotinic acetylcholine receptor, a molecular assemblage found throughout the animal kingdom. It is an essential constituent of the human brain as well as of the jellyfish nervous system, whose ancestral lines may be over a billion years old (Smith 1996; Dawkins 2004). Although the human mind is sometimes viewed as evolution's greatest achievement, its primary neurological components are based on very old technologies. Recognizing the constraints of evolution will allow us to understand how something as quirky as schizophrenia could be evolutionarily advantageous.

The task of deciphering the inner workings of any biological organism is aided by knowing the ground rules or tendencies of evolution. The first section of this chapter will review a few basic evolutionary mechanisms, particularly those that may relate to shamanism. Section two concentrates on the field of ethology, the scientific discipline that relates to animal behavior. Animal behaviors turn out to be connected to human social customs because certain kinds of social behavior are repeatedly observed throughout the animal kingdom, and these archetypal arrangements have probably guided the progression of human cultural practices. Section three investigates a few important contributions from the new field of evolutionary psychiatry which may elucidate the origins of various psychiatric disorders. The last section is devoted to anthropology — specifically, those aspects of ancient traditional life that seem to be most relevant to shamanism. Such an examination is essential because it is the selection pressures associated with hunting and gathering societies that placed the finishing touches on the modern human mind.

Basic Evolutionary Theory

Much has been written about Darwin and his theory of natural selection. The basic idea begins with a simple fact obvious to everyone: that organisms in a species tend to be slightly different from each other. In other words, nature is filled with variation. Darwin surmised that part of this variation was innate — not necessarily the result of environmental effects. Darwin's next observation was that organisms have the ability to produce many copies of themselves, but because exponential growth is impossible (on average only one progeny, or two per couple, can survive in sexual species), some type of culling had to occur.[1] Darwin figured that the process of controlling the population of a species was not random, but rather some organisms were better equipped to survive and reproduce than others. The consequence of these axiomatic observations is that some sort of competition exists within species. Darwin called this competition *natural selection*. (Herbert Spencer later coined the phrase "survival of the fittest.") Gradually, over many generations, natural selection changes species, one increment at a time. Darwin called this, "descent with modification," but we now refer to the entire process as evolution (Darwin 1859).

Evolution requires three components. The first is replication, but a form of replication that results in slightly inexact copies. This subpar fidelity produces the second element of evolution: variation. The third and final requirement is competition between these slightly variable forms. This three-step scheme is Darwin's essential insight. It is so simple that one of Darwin's earliest proponents, T. H. Huxley, was compelled to comment, "How extremely stupid not to have thought of that."

The Origin of The Origin of Species

In retrospect, there are several reasons why evolution was not so obvious before the nineteenth century. First, the idea that the earth was ancient — much older than the biblical estimate of about 5000 years — was just beginning to take hold in scientific circles. Charles Lyell's *Principles of Geology*, published in the early 1830s and which Darwin read during his voyage on the *Beagle*, concluded that the earth was far older than a mere few thousand years. Lyell's book made possible Darwin's concept of gradualism: the idea that very small differences stretched over extremely long timescales can generate momentous changes — analogous to how running water carves magnificent canyons.

Another quality that likely allowed Darwin to understand evolution before anyone else was his atheism.[2] Religion tends to exalt humankind through our supposed special relationship with God. The *scala naturae*

---

[1] This idea was cemented for Darwin after reading Thomas Malthus's "An Essay on the Principle of Population" (1798).
[2] Although careful not to advertise it too vociferously, Darwin's philosophy was closer to atheism than agnosticism.

or great chain of being is an ancient Western formulation that arranges the universe in a hierarchy ostensibly so obvious that few even thought to question it before Darwin's time: God at the top, followed in succession by angels, demons, man (further classed as kings, princes, noblemen, etc.), animals, plants and finally different forms of rock. Darwin's atheism allowed him to readily scrap humankind's position in this supposed hierarchy of nature. Although admittedly some organisms may be more complex than others, we are all equal when it comes to the main purpose of evolution: survival. Evolution was therefore counter-intuitive to the nineteenth century mind because it connected man to the rest of nature in a very mundane way.

Although zoology has, perhaps perforce, readily accepted the tenets of evolution, other scientific fields have typically been far more resistant to the implications of Darwin's theory. Daniel Dennett, author of *Darwin's Dangerous Idea*, compares evolution to an imaginary substance he calls universal acid: "it eats through just about every traditional concept, and leaves in its wake a revolutionized world-view, with most of the old landmarks still recognizable, but transformed in fundamental ways" (Dennett 1995). The old-timers in each academic sect have historically tried to contain or wall off the idea, "out of cosmology, out of psychology, out of human culture, out of ethics, politics and religion" (Dennett 1995). For example, the brilliant and influential linguist Noam Chomsky has consistently downplayed evolution's role in the development of language, even though there is compelling evidence that language is the primary determinant behind humankind's uniquely over-sized brain (Deacon 1997).[3]

Like poker or chess, the basic rules of evolution are simple but the game is infinitely more complex. To investigate the central strategies of natural selection, it will be useful to divide our analysis into evolution's three elemental parts: replication, variation, and competition. Our focus will be on those aspects of evolutionary theory with the greatest relevance to psychiatric disorders.

---

[3] It cannot be ignored that the human brain is an energy hog, consuming about 20 percent of our food intake. Therefore its primary abilities must be evolutionarily advantageous. I am partial to Terrence Deacon's idea that *Homo sapiens* is nature's "symbolic species," language having evolved during a period ranging from as little as several hundred thousand years to perhaps as long ago as a few million years. Most cognitive skills such as calculation, grammatical rules and theory of mind appear to be based on fairly straightforward algorithmic rules and as such are perhaps only modest extensions of the cognitive abilities typically observed in animals. In contrast, symbols (i.e., words) are each rooted in a large complex web of memories, ideas and other associations that are accessed and sequenced (i.e., into sentences) at extremely high speeds. Processing symbols seems to require high-powered computation and as a consequence, significantly increased brain mass. Our ability to reflect and think abstractly may be a fortuitous but unintended consequence of the sophisticated cognitive hardware already in place.

Replication

In 1859, when Darwin first published *On the Origin of Species*, the inner workings of biological replication were essentially a black box (Darwin 1859). In 1865, Gregor Mendel discovered that the inheritance of certain physical traits in pea plants followed specific rules. It turns out that these basic rules of inheritance are followed by all of nature, and Mendel's heritable "factors" corresponded to genes — the fundamental units of heredity. However, the greater scientific community was not aware of Mendel's scientific paper until the early 1900s.[4] The next great discovery in the field of genetics occurred in 1953, when James Watson and Francis Crick revealed the unique molecular structure of these replicating factors: the double helix structure of DNA (Watson 1968).

Although some disorders with psychiatric consequences, such as Down syndrome, are the result of problems during chromosomal replication, there is no appreciable evidence that schizophrenia reflects a dysfunction in the genetic replicative process.

Variation

Variation is a measure of the extent of change, typically within a sample of related units (i.e., a set of points) (Robinson and Schluter 2000, 72). Part of the grandeur of nature relates to its variation. The extent of this variation is amazing, especially when one considers that evolution tends, with some exceptions, to move towards a single ideal phenotype. Despite 150 years of evolutionary research, the study of biological variation is, remarkably, still in its infancy. Consequently, the fundamental determinants of variation are still being worked out (Hallgrimsson and Hall 2005; Mousseau et al. 2000; Weiss 1993; Wool 2006).

Evolution requires nature's components (DNA and organisms) to produce inexact copies. If the earliest primordial DNA had reproduced itself with 100 percent fidelity, nature's diversity would not have been possible. The replicative shortcomings of DNA are fortunate, and principally dictated by the rules of molecular biochemistry. However, in addition to these elementary biochemical laws, nature seems to have found ways to expand or diminish the molecular set point of variation. In other words, nature appears to massage both DNA (genotype) and individual organisms (phenotype) to achieve optimum variation (Jablonka and Lamb 2006, 86). There may be certain environmental clues that prompt organisms to accelerate

---

[4]Mendel had certainly read *On the Origin of Species*; however, it appears that Darwin had not come across Mendel's paper (Galton 2009).

(i.e., increase mutation rates) or put the brakes on variation (e.g., genetic canalization) — in effect hedging one's bets in nature's competition.[5]

The amount of variation corresponding to any ideal phenotype deviates around a Gaussian mean in its own unique way. How much variation is to be expected? Some traits show hardly any disparity (numbers of fingers) while others demonstrate greater variability (height, breast size). Evolutionary theorists do not yet know precisely how the limits of variation are determined.

How about social behaviors such as aggressiveness, mating interest, obsessive-compulsivity, or parenting style? How far can instinctive behaviors deviate before they find themselves on the evolutionary radar as a disadvantageous trait? Although such questions cannot yet be answered, it would seem important for psychiatrists — who regularly deal with "deviant" phenotypes — to pay greater attention to the science of evolutionary variation.

Biological variation can be observed through nature's organisms — phenotypic variation — or its diverse strands of DNA — genotypic variation. The study of biological variation is exceedingly complex because it mixes the statistical conundrums of population studies — phenotypes — with the intricacies of molecular genetics — genotypes. Not all genotypic variation results in phenotypic variation since varied genes can sometimes result in identical phenotypes. On the flip side, because certain environmental conditions can turn genes on or off, phenotypic variation can occur with identical genotypes. Although they are interlinked, it will be practical to review genotypic and phenotypic variation separately.

Genotypic variation

The ultimate creator of both genotypic and phenotypic variation is genetic mutation. There are several ways to bungle a copy of DNA (Freeman et al. 2006):

  i. Single nucleotide polymorphisms (SNPs)
 ii. Variable number of tandem repeats (VNTRs) (also known as microsatellites and minisatellites)
iii. Transposable genetic elements (transposons)
 iv. Polyploidy
  v. Structural alterations in DNA — deletions, duplications, inversions, translocations

This latter group can result in DNA copy number variation (CNV), briefly mentioned in Chapter 4. Until recently, it was believed most genetic

---

[5]Canalization is the accumulation of genetic variation without phenotypic expression. The changes in the phenotype conveniently emerge during periods of stress to the organism. It is a theoretical genetic process with some preliminary experimental evidence to support it.

variation resulted from SNPs, but it now appears that CNVs are probably responsible for the lion's share of genetic variation in humans. There has been some recent excitement among psychiatric researchers regarding CNVs, but whether this form of genotypic variation has anything to do with the major psychiatric disorders awaits further research. (Fanciulli et al. 2010; Ikeda et al. 2010; Kirov et al. 2009; St Clair 2009; Sutrala et al. 2007; Tam et al. 2009).

Sex may be another modifier of genetic variation (Ridley 2001). Evolutionary theorists do not entirely agree why sex even exists in nature, although sexual organisms do seem to experience a number of auxiliary evolutionary benefits. Sexual recombination of genes enhances genotypic variation by shuffling genes at every generation. This intensified random intermixing seems to hasten the aggregation of cooperative genes, while eliminating clusters of undesirable mutations.

Another source of genetic variation may be undesirable environments, which under certain circumstances may accelerate mutation rates (and chromosomal recombination rates) (Jablonka and Lamb 2006, 262, 284). Such stress-induced mutations have been identified in lower organisms like bacteria and fungi; whether they also apply to animals with longer intergenerational periods remains to be seen. (This type of mutation should not be confused with the stress diathesis model of schizophrenia, which proposes that environmental stress acts on a preexisting vulnerability to trigger schizophrenia. This model does not directly involve somatic or germ line mutations but rather epigenetic changes. In the stress diathesis model of schizophrenia, environmental stress could conceivably alter schizophrenic gene expression (but not the actual genes) through DNA methylation and other biochemical modes of histone modification.)

Phenotypic variation

Richard Dawkins's classic book *The Selfish Gene*, published in 1976, heralded the gene-centered perspective of evolution. According to Dawkins, genes are the primary players in the evolution game, while organisms are mere vessels created by genes to ensure their survival — talk about the ultimate cosmic downgrading of humankind. However, like a tank with a soldier inside, it is the vessel — the organism's phenotype — that finds itself in a fight to the death with other organisms. Therefore, the evolutionary principles behind phenotypic variation are of utmost importance.

In theory, natural selection tends to choose the most ideal phenotype for any given environmental situation and correspondingly, the single best gene supporting a particular trait will eventually supplant all others (Ridley 1996). However, there are many more alternative genes floating around in nature — presumably these genes are the second- or third-best options for any particular trait. Take eye color: which color is evolutionarily superior? If mating is otherwise random, a tiny preference for one eye color over another, over many generations, will result in the superior eye color displacing all others. But how do genes that are not necessarily superior to the supposed front-running gene increase in prevalence? There is a lot more variety in nature than one would predict using the basic rules of evolution (and indeed, evolutionary theorists are beginning to reveal many auxiliary rules of evolution).

A species that demonstrates variable phenotypes for a particular trait is said to possess a polymorphism. A polymorphism is the presence of two or more distinctive genes (i.e., alleles) at a certain position (genetic locus) in the DNA. A polymorphic allele must be present at a frequency above 1 percent in the population. (This is an arbitrary cutoff to disregard alleles that are clearly disadvantageous mutations.) In other words, if an alternative gene is present at a frequency above 1 percent, there's a good chance it is not a completely incompetent gene, since it must have spread at some point in the species' history.

The question, "How does a genetically based psychiatric disorder come to be?" is equivalent to, "What maintains phenotypic variation?" which can in turn be reframed as, "What maintains the genetic polymorphism?" Recall from Chapter 4 that, due to those candidate schizophrenia genes, schizophrenia is presumably the result of several polymorphic genotypes.

It should be underscored that the majority of psychiatric problems may not be the result of significantly disparate genotypes. For example, clinical depression in Western societies is usually connected to a loss of attachment or reduction in status — and most humans are vulnerable to depression if the circumstances become severe enough.[6] Similarly, a diagnosis such as borderline personality disorder may conceivably be connected to some sort of genetic vulnerability that wasn't very important prior to modern times. The emergence of borderline personality disorder almost always requires substantial childhood neglect or abuse (Bandelow et al. 2005; Zanarini et al. 1997), which may have been rare in the communal social structure of traditional societies. In other words, the characteristics which must exist in order to apply evolutionary principles to a psychiatric problem — that is, a genetically distinct variable form with reduced fecundity and long-standing selection pressures, which seem to exist for schizophrenia — may not necessarily be present for the majority of emotional ailments.

Although the question, "What maintains a genetic polymorphism?" hasn't been a pressing one for clinical psychiatrists, it could be one of the most fundamental problems in the study of schizophrenia (as well as OCD,

---

[6] The DSM-IV purposely avoids categorizing the possible subtypes of depression — not because etiological differences aren't suspected but because the field cannot agree on how to divide them.

bipolar and autism). In a seminal paper published in 2006, evolutionary scientists Matthew Keller and Geoffrey Miller (neither is a psychiatrist) challenged the psychiatric community about this very point. They argued that most psychiatric disorders (I would argue only schizophrenia, bipolar, OCD and autism) must fit into one of a short list of established categories that explain persistent variation in a population. Although some evolutionary psychiatrists had already thought about the problem, Keller and Miller's paper methodically framed it better than any predecessor: simply, how do we explain the supposed stable genotypic variation characteristic of certain psychiatric disorders? They listed a number of evolutionary mechanisms to explain the existence of polymorphisms. I have arranged the possible types of variation slightly differently without affecting their central arguments. Perusing the theoretical evolution literature, I have identified eight possible ways to explain genetic polymorphisms — most of which do not seem to apply to schizophrenia.

1) *Genetic drift* is an established idea that some polymorphisms exist only by random chance. It is possible that different alleles can haphazardly increase or diminish in frequency, especially if the phenotypic effects of the alternate alleles are very similar to the primary alleles (i.e., neutral fitness). Genetic drift is more apt to occur in smaller populations. It is not believed to be a factor in psychiatric conditions (Keller and Miller 2006).

2) *Directional evolution in transit* (Wool 2006, 132). Sometimes polymorphisms are observed because we are actually catching evolution at that cross section in time when a fitter gene is in the process of replacing another one. This is rare and also not believed to be the primary explanation of psychiatric conditions.

3) *Mutation-selection balance* represents the presence of disadvantageous genes due to a recent mutation; nature has not yet had time to select these particular alleles out. This process could explain the presence of very rare genes but is not likely to explain a true polymorphism (genes in the frequency of greater than 1 percent at a particular genetic locus).

4) *Polygenic mutation-selection balance theory*: This is Keller and Miller's own theory: although any given single mutation has a low frequency, all of the genes that contribute to the brain and behavior — perhaps hundreds — could cumulatively add up and account for the present rate of mental illness. It is a clever idea but in my view, it doesn't satisfactorily explain how hundreds of haphazard mutations channel their phenotypes into a shortlist of specialized psychiatric syndromes. Schizophrenia is not represented by a boundless array of slapdash problems but rather a reoccurring pattern of distinguishing features.

5) *Negative frequency-dependent selection*: It is believed that the simple act of being rare can sometimes increase an organism's fitness (Hori 1993). This process could maintain scarce genes at a fixed low-frequency equilib-

rium. It is like flying under the radar of selection pressures. For example, an edible species of butterfly may contain a small percentage of individuals whose wing patterns resemble a poisonous species. This minority phenotype will be left alone by predatory birds and consequently prosper — but only to a point. As the numbers of imposters increases, predators may not so rigorously avoid them (Ridley 1996, 122). Keller and Miller suggest that sociopathy may be a human trait that is tolerated when rare but as it increases people take active steps to avert it (e.g., locking their doors).

It has also been suggested that alternative mating strategies can be maintained by negative frequency-dependent selection. Nature provides several examples of two different heritable mating strategies existing among the males (or females) of a species. For example, while most bluegill sunfish males are large and fertilize their chosen mate's eggs with her approval, smaller male "sneaker" bluegills swim under the preoccupied couple and release their sperm covertly. The uncommon form survives because it possesses certain advantages directly related to its rarity (Rios-Cardenas et al. 2007; Gross 1991). Schizophrenia could conceivably be such a strategy. Recall Nettle and Clegg's proposition that the schizotypal phenotype is propped up by creativity and its relationship to mating success. Schizotypal creativity could make people unique in ways that make them a tad more attractive.[7]

6) *Temporal-Spatial Variation Selection* refers to differences in geography or other aspects of the environment that can alter the adaptive profile of various genes. For example, the behavioral trait of shyness (versus boldness) often seen in fish is preferred in certain environments but not in others: shyness is more adaptive during periods or in places of high predator density, while boldness is more evolutionarily effective during periods of low predator density. Nettle (2006b), in a thoughtful paper, suggests that personality traits such as extroversion and introversion could be characterized by genetic polymorphisms under this sort of influence. For example, extroversion may be risky during times of war but more adaptive during peacetime.

7) *Antagonistic pleiotropy* refers to the theoretical situation where a singular gene affects more than one phenotypic trait — one trait is beneficial while others may be disadvantageous, but not enough to offset the advantage (Weiss 1993, 167; Keller and Miller 2006; Burns 2009). I agree with Keller and Miller, who suggest that antagonistic pleiotropy is not likely to be an evolutionary mechanism that maintains genetic variation, because whichever allele has the best ratio of advantage to disadvantage will eventually supplant all others, thus erasing the polymorphism. How-

---

[7]Negative frequency-dependent selection can work either synergistically or independent of group selection.

ever, antagonistic pleiotropy could, in theory, explain the presence of disadvantageous behavioral traits stemming from one predominant gene.

Two evolutionary psychiatrists, Jonathan Burns and Randolph Nesse, appear to be thinking along the same lines. They suggest that schizophrenia represents the downside of valuable, yet-undetermined neurobehavioral genes. Burns speculates that antagonistic pleiotropy (along with gene-environment interactions) could apply to schizophrenia (Burns 2009). Nesse is more circumspect about the direct application of antagonistic pleiotropy to schizophrenia. Instead, he uses the similar but arguably less well-defined notion of "evolutionary tradeoff" (Nesse 2004). Nesse uses a very straightforward example to explain his hypothesis: gout. Elevated uric acid levels may increase evolutionary fitness by protecting against oxidative tissue damage, but too much uric acid causes gout.

It is certainly plausible that this dynamic could be applicable to schizophrenia and other psychiatric disorders; however, I do not believe it can explain the whole story. If we assume that mental illness reflects a veritable disease process, we must acknowledge that no other tradeoff in nature is built as shoddily as the human neurobehavioral system. In fact, I would argue there is no precedent in nature. For example, gout's prevalence is appreciably less than schizophrenia's 1 percent figure (and it was probably even more rare in traditional societies due to lower-protein diets), and gout doesn't reduce fecundity as drastically as schizophrenia.

8) *Heterozygote advantage* is the name for a situation when two or more alleles assort themselves in a population according to specific ratios, based on the relative fitness of each allelic combination. Heterozygote advantage maintains a balanced polymorphism because the organism with two disparate alleles at a particular gene locus (the heterozygote) holds an evolutionary competitive edge over both homozygotes. Sickle cell anemia is perhaps the best example of heterozygote advantage in human beings. In malarial zones, having two normal alleles is disadvantageous because it makes you too susceptible to malaria, but having two sickle cell alleles makes you too vulnerable to cardiovascular problems. Having one sickle cell allele in combination with a normal allele is an excellent compromise and therefore the most evolutionary advantageous position. In my view, heterozygote advantage holds the greatest promise to explain schizophrenia and other genetically based psychiatric disorders.

According to one textbook about genetic processes in populations, "heterozygotes, with their two alternative genetic factors at their loci, may be more 'flexible' physiologically since they may produce more kinds of enzymes" (Wool 2006, 80). It has long been observed that in nature, hybrids usually fare better than their homozygotic counterparts. This is called *heterosis* or hybrid vigor and is seen most noticeably in plants (e.g., maize, rice) (Xiao et al. 1995; Birchler et al. 2003).

Although hybrid vigor is often casually observed, to definitively prove heterozygote advantage is not so easily done (Gemmel and Slate 2006). Here is the catch-22 of identifying heterozygote advantage: it is very difficult to prove heterozygote advantage when homozygotes are not lethal, but this is precisely the situation where heterozygote advantage is most conspicuous. In other words, heterozygote advantage becomes more feasible (but less obvious) as we narrow the fitness gap between the heterozygote and the homozygote. It is possible that the more benign forms of heterozygote advantage are widespread, which could explain why so many polymorphisms exist in nature.

Keller and Miller have two reservations about heterozygote advantage that are not unreasonable, but dubiously applicable. They claim that heterozygote advantage is usually an evolutionary stopgap measure (e.g., to contend with cataclysmic threats such as virulent infections), and that there are only a few unequivocal examples in nature. (They claim six examples based on Endler's 1986 manuscript.) These two claims may be true for classic cases of heterozygote advantage, but such cases are nature's extreme cases — when the homozygote is so crippled that it contains lethal genes, like in cystic fibrosis (Bertranpetit and Calafell 1996; Dean et al. 2002; Schroeder et al. 1995) or sickle cell anemia (Allison 1954; Ashley-Koch et al. 2000). Less explicit cases, where homozygotes are just slightly less adaptive, may be much more common in nature. The traditional examples of heterozygote advantage are probably just the tip of the iceberg. In fact, more examples than Endler's six examples cited twenty-five years ago are beginning to come to light (Gemmel and Slate 2006).

Competition

We have thus far reviewed two of the three cornerstones of evolution: replication and variation. We will now explore the science of how organisms compete with each other and survive to the next generation.

Evolutionary adaptation functions primarily at the level of the individual. Individuals compete with each other, with those being most "fit" surviving and populating the next generation. However, evolutionary competition can also occur between groups. Admittedly, group selection (also known as multilevel or interdemic selection) tends to be less conspicuous and widespread than individual selection.

Group competition can theoretically occur between sexual pairs, kin groups (families), and lesser-related groups (honeybee colonies, wolf packs, hominid tribes). There may also be competition at the species or phylum level (Gould 2002). The science of multi-level selection will be reviewed in detail because it may explain the evolutionary origins of shamanism.

Group selection

Group selection has historically been a controversial idea (Borello 2005). Any discussion of the importance of group selection in any given evolutionary situation was usually followed by reasonable debate — that is, until the 1960s, when the critics of group selection won the battle in the scientific literature. From then until about fifteen years ago, most biologists completely discounted the idea. Although the fiercest objectors have relented, in my view group selection continues to be under-appreciated. It should be noted that no one disputes that the forces of individual selection overwhelmingly steer evolution. The argument centers on whether group selection is of secondary importance or is virtually insignificant.

The origins of the group selection debate began with a vexing question: "How can unequivocally disadvantageous traits, like altruism, exist?" For some, group selection was a satisfactory answer. Others, however, invented explanations that seemed to avoid group selection dynamics: kin selection (which doesn't explain altruism between unrelated members of the same species) or reciprocal altruism (which does not explain non-reciprocal altruism).

It is indisputable that the concerns of the group selection skeptics were not trivial. Group selection represents the propagation of traits that enhance group survival but are either neutral or act *disadvantageously* upon the individual. Many theorists intuitively felt that going against the momentum of evolution's primary propeller, individual selection, was unsustainable for long periods — a legitimate concern. However, sensible reservations turned into dogma.

I have often wondered whether resistance to group selection is related to scientists not playing enough team sports (Buchanan 2000). Every observant athlete knows that games can be won or lost over the subtlest disposition of a few teammates. In hockey, for example, a defenseman who unwittingly lines up closer to his defensive counterpart (the equivalent of a neutral mutation) can, by stumbling upon a better defensive formation, determine the outcome of a game. Small positional differences that are neither advantageous nor disadvantageous to the individual (i.e., it won't change how much the player scores) can reap dividends for the group. Using evolutionary parlance, neutral traits can eventually become long-standing vestigial traits if there is no better phenotypic alternative. This, for example, explains the existence of many vestigial traits in humans: the coccyx, male nipples, the appendix. Now, imagine that the vestigial trait helps the group — that's group selection. Notice that we haven't even yet had to introduce the concept of heterozygote advantage, a mechanism that does exist in nature and can support even substantially disadvantageous (i.e., altruistic) traits!

The original advocate of group selection was Charles Darwin. In *The Origins of Species*, Darwin discusses honeybees and ants, referring to them as "social insects" (1859, 182). To explain the "sterile condition of certain members of the community," Darwin invoked the concept of group selection. He rationalized, "We can see how useful their production may have been to a social community of insects, on the same principle that the division of labour is useful to civilized man" (p. 185). Darwin also felt that group selection applied to *Homo sapiens*, "At all times throughout the world tribes have supplanted other tribes..." (1871).

The next notable contribution to the theory of group selection came from Peter Kropotkin who in 1902 published "Mutual Aid: a factor in evolution." Kropotkin acknowledged the primacy of individual selection but felt that instances of cooperation in nature were being overlooked due to the new emphasis on Darwinian competition. Born a Russian prince, Kropotkin renounced his noble titles and advocated for a form of anarchistic communism. His interest in group selection may have been rooted in his appreciation of the political troubles of that period. Kropotkin understood that Russia's tyrannical feudal system was effectively turning it into a failed state, in large part due to peasant resentment and a consequent lack of genuine cooperation. Having studied zoology as well as having trekked through Siberia on several scientific expeditions, Kropotkin recounted numerous examples of cooperation in various species such as ants, parrots, and wolves. Kropotkin surmised that cooperative behavior in nature was fostered by the forces of evolution — specifically, group selection (Borello 2004).

In 1962, V. C. Wynne-Edwards published *Animal Dispersion in Relation to Social Behaviour*, which contained numerous ethological observations in support of group selection. Wynne-Edwards's logic was as follows (p. 19):

> In the *extreme* case [my emphasis], in various social insects, it has been possible to evolve castes of sterile individuals, something that is inconceivable in a world where the most successfully fecund were bound to be individually favoured by selection and the infertile condemned to extinction; it could only have evolved where selection had promoted the interests of the social group, as an evolutionary unit in its own right.

In other words, if group selection can produce genetically sterile organisms — the most disadvantageous trait imaginable for an organism — then it shouldn't be much of a stretch to find other examples where minor phenotypic hindrances are maintained by group selection. Wynne-Edwards provided six-hundred-plus pages of candidate group selection traits, ranging from population regulation in fish to synchronized activities in birds.

But Wynne-Edwards's magnum opus on group selection was villainized in the evolutionary literature as the product of naive and sloppy thinking. He was unfairly criticized for not being an evolutionary theoretician — *Animal Dispersion* is clearly a naturalist's work and not a theoretical analysis of group selection. However, it was disingenuous to expect Wynne-Edwards to "prove" that, for example, population regulation was unequivocally related to group selection. There is no algorithmic way to prove that even the most obvious phenotypic adaptation is truly evolutionarily beneficial. Nature is so immensely complex that there is always room for one more caveat or alternate explanation, and the evolutionary adaptiveness of any given trait is ultimately an arbitrary determination. In the end, Wynne-Edwards's fiercest theoretical critics, like G. C. Williams (1966), seem to have got it wrong while the naturalistic observations in *Animal Dispersion* continue to be compelling examples of group-selected traits.

The legitimacy of group selection received a significant boost after Wilson and Sober's (1994) seminal paper, "Reintroducing group selection to the human behavioral sciences." This paper made two important contributions. First, the Wilson and Sober article became the vanguard of a new wave of research (Goodnight and Stevens 1997; Wade 1976; Wade 1977; Wade 1978) clarifying the experimental and theoretical underpinnings of multilevel selection — and appeasing some of the critics (Sober and Wilson 1998). Second, Wilson and Sober "re-established" group selection as a plausible evolutionary mechanism to explain human altruism. Applying the principle of group selection to human populations has historically been controversial. Darwin posited group selection to explain the origins of morality, but most subsequent evolutionary theorists have balked at this conceptual extension (Darwin 1871; Sober and Wilson 1998, 4).

As researchers become attuned to the possibilities of multi-level selection, more attention is being given to its theoretical basis, as well as its potential role in the evolution of hominoid tribes. Professor Martin Nowak and colleagues at Harvard have investigated the mathematical underpinnings of multi-level selection and extended the theoretical framework supporting this evolutionary phenomenon (Nowak 2011). Their formulas suggest that group selection works better when:

- larger numbers of groups compete;
- migration is minimized;
- there is greater social delineation of the group.

All of these features are typical of human social groups: tribes, teams, nations, corporations. Another evolutionary scientist, Samuel Bowles, has shown in mathematical terms that increasing the lethal effects of intergroup competition (i.e., intertribal war) preferentially favors altruistic (self-sacrificing) genes (Bowles 2006).

There are a number of candidate group-selected traits, of which three in particular have a special connection to the psychological sciences: altruism, complex communication and task specialization. These traits are more readily observed in animals that compete in discrete groups, and are conspicuously absent in solitary creatures. Notice that complex communication and task specialization are arguably special cases of altruism: complex communication is disadvantageous because, if it is honest, it reveals one's motives, while task specialists often sacrifice their fecundity to provide exclusive services for the group. It should also be noted that certain altruistic behaviors including complex communication can be partially explained by kin selection or reciprocity paradigms, and that such mechanisms may work in concert with group selection. Indeed, maternal altruism may be the original template for group-selected altruism. In other words, kin selection may sometimes be a prerequisite for group selection. The success of task specialization, however, seems to rely primarily on the dynamics of multi-level selection.

The species in which group selection is most obvious are ants, bees, wasps and termites, collectively known as the eusocial insects (Choe and Crespi 1997). The level of social cooperation in these creatures is astounding, and consequently colonies take on the characteristic of a self-sustaining organism, sometimes described as a superorganism (Hölldobler and Wilson 1994). All of the signature group-selected traits can be found among the eusocial insects: altruism, complex communication and task specialization.

Altruism is widespread in the eusocial insects and takes on several forms. For example, soldier ants or guardian bees will defend against invaders at great risk to themselves. In certain ant species, raids against neighboring colonies are so highly organized that one can't help but make a comparison to traditional human warfare. Perhaps the greatest altruistic sacrifice is sterility, which is found in certain termite and honeybee worker castes.

Complex communication is perhaps best exemplified by the waggle dance in honeybees, a form of communication unequaled in complexity among insects and only rivaled by a few social mammals (Von Frisch 1953; Seeley 1997). Ants use pheromones, sounds and body movements to communicate such expressions as "attraction, recruitment, alarm, identification of other castes, recognition of the larvae and other life stages, and discrimination between nestmates and strangers" (Hölldobler and

Wilson 1994, 55).

Of all the species in the animal kingdom, task specialization is most easily identified in ants and bees. Specialists can be found in other clannish species (i.e., animals that cluster with unrelated conspecifics) but such roles are not usually as distinct. In contrast, specialized roles are never observed in solitary species such as spiders, polar bears or foxes. Like the division of labor that makes factories more efficient, task specialization helps certain insect colonies outcompete others. Task specialization is characterized by extreme phenotypic variability between members of that species, in order to do their varied roles (Robinson 1992). In general, extreme phenotypic variability is rare between members of the same species because uniform selection pressures tend to homogenize individuals; ideal traits become preferentially selected and eventually become ubiquitous. Leafcutter ants, in contrast, are comprised of several task specialists such as foraging workers, gardeners and soldier ants (Hölldobler and Wilson 1994). Some soldier ants can be up to 300 times heavier than gardener workers of the same species (Hölldobler and Wilson 1994). Preliminary experiments seem to confirm that eusocial insect castes are determined, in part, by genetic factors (Breed et al. 1990; Dreller 1998; Oldroyd et al. 1994; Page and Robinson 1991; Anderson et al. 2008; Kitade et al. 2010).

Group-selected traits are not only observed in eusocial insects but also among social animals, including human beings (Dugatkin 1999). One common altruistic activity is sharing. However, sharing may not always be genuinely altruistic — it can be only considered altruistic when it is non-reciprocal and occurs between unrelated individuals. Veritable altruistic sharing occasionally occurs in, for example, chimpanzees and vampire bats (Wade 2006; de Waal 1996).

Organized warfare and its various ancillary forms, such as mobbing, raiding and cooperative hunting, represents another type of altruistic behavior. Mobbing — ganging up to drive away a larger attacker — is a communal behavior occasionally seen in some bird species (Dugatkin 1997). Cooperative hunting is the norm for dolphins, lions and chimpanzees. Small groups of male chimpanzees, while patrolling their territory, have been known to raid and kill chimpanzees from other bands (Goodall 1986; Dugatkin 1997). Old age and death may be the ultimate altruistic activity — a potentially group-selected genotypic trait at the level of a species or phylum (Mitteldorf 2006).

Complex communication is another phenotypic trait typically observed in social animals and probably fostered by the forces of group selection. It seems that elaborate signaling could serve any familial relationship (e.g., the parent-child relationship); however, complex communication tends to be found only in animals that spend a lot of time with unrelated conspecifics. The most sophisticated communication is found in such clannish species as honeybees, dolphins, whales and primates.[8] In dolphins and whales, there are frequent exchanges of diverse underwater sounds. Although clicks are used for navigation, an assortment of whistles and burst-pulse sounds communicate various aspects of dolphin social life (Bearzi and Stanford 2008). Most primate species possess an array of alarm calls. Chimpanzees have the most, at about fifty different calls (Bearzi and Stanford 2008, 174).

Human language is arguably nature's greatest evolutionary invention and the pinnacle of complex signaling. Other possible forms of group-selected communication are humor, music, dance and rhyming (to improve a tribe's ability to remember cultural events). It has also been forcefully argued that religion is both a type of communication (Steadman and Palmer 2008) and a group-selected altruistic trait (Wilson 2002).

Another group-selected trait, task specialization, can also be found in social animals, however compared to eusocial insects, mammalian specialization tends to be less defined. One of the best examples of task specialization among mammals is the naked mole rat, a burrowing rodent indigenous to East Africa, whose colonies of fifty to a hundred individuals include a caste of sterile female workers (Faulkes and Bennet 2001; O'Riain et al. 2000). The distribution of their food source makes it obvious why group cooperation is so important: these rodents fan out underground in search of rare underground tubers that are often large enough to feed an entire colony for months (Judd and Sherman 1996). This kind of cooperation appears to be the most efficient solution under such circumstances. It is not yet known how or whether genetic factors or environmental cues (or a combination of both) determine task specialization among naked mole rats.

The use of rotating sentries seen in dolphins or Canada geese can be considered a diluted form of task specialization (Norris et al. 1994). In tamarin monkeys, which live in cohesive groups of five to ten members, a single individual, usually the dominant adult male, performs all the sentry duties (Garber and Bicca-Marques 2002). Social hierarchies may also be a form of task specialization (Figure 6.1). Most evident in such species as wolves, hyenas and primates, hierarchal social structures can mitigate disputes within groups by assigning specific social roles as determined by a pecking order (Cheney and Seyfarth 1990; Polimeni et al. 2005).

I have provided a detailed explanation of group selection because of

---

[8]Mating songs in crickets or birds, for example, are another example of complex communication; perhaps at the level of a dyadic mating couple (i.e., a group of two). Birds that form long-term monogamous pairs are more apt to learn duets (Pepperberg 1999, 324). This is in accordance with multi-level selection theory, because there would be greater selection pressures to acquire communication skills in birds that survive as a dyadic unit.

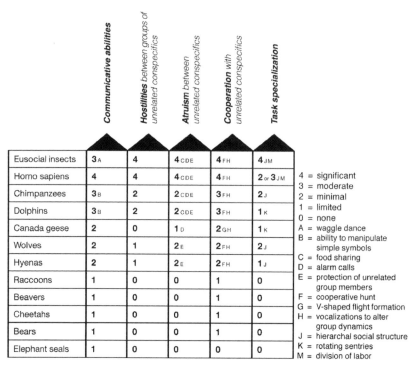

Figure 6.1: Group selected traits in a variety of animals. Eusocial insects, humans, chimpanzees, dolphins and Canada geese commonly aggregate with non-relatives. Wolves and hyenas assemble in clans of mixed family members and non-relatives. Raccoons and beavers live in family units. Cheetahs are usually solitary, although males occasionally cooperatively hunt with non-relatives. Bears and elephant seals are typically solitary.

its role in the shamanistic theory of schizophrenia, which claims that shamans were task specialists in hominid tribes. This argument, which I will develop in later chapters, is based on a number of observations including the distinctness of the shaman-schizophrenia phenotype, the heritability of schizophrenia (and presumably shamanism too) and the universality of shamanism and schizophrenia throughout the world.

Although the genetic mechanisms behind group selection have not been conclusively established, it has been proposed that group-selected phenotypes could be maintained by either, or both, of two types of balanced polymorphism: heterozygote advantage and assortative mating. We have already seen how heterozygote advantage can lead to the maintenance of disadvantageous phenotypic traits which could be used in the service of altruism, complex communication or task specialization. Assortative mating refers to the tendency for individuals to choose mates that possess one or more heritable traits that the individual themself possesses. This creates non-random pairings that increase variation within a species. For example, if tall people prefer each other, their children will usually be a lot taller than average, thereby increasing variation in the community.[9]

Thus, assortative mating is yet another possible mechanism for phenotypic variation (Wilson and Dugatkin 1997). Wilson and Dugatkin conclude, "Assortative interactions can generate highly nonrandom variation among groups, favoring the evolution of altruism and other group level adaptations among genealogically unrelated individuals." In Chapter 11 I will examine the possibility that assortative interactions are relevant in schizophrenia.

Ethology

Twenty years ago, having just graduated from psychiatry residency, I picked up a copy of Jane Goodall's book *Through a Window* (1990), relieved to finally read something having nothing to do with psychiatry — at least, that's what I thought. The book recounts Goodall's observations of the wild chimpanzees of Gombe, Tanzania. In one part of the book, Goodall tells the story of Flint, an eight-year-old adolescent chimpanzee who becomes stricken with extreme grief after losing his mother (Goodall 1990, 196–197). Goodall described the animal as looking depressed, showing such behaviors as lethargy, social isolation and refusal to eat — all of which are familiar neurovegetative signs of depression. Two weeks after his mother passed away, so did Flint. The last common ancestor of chimpanzees and human beings is believed to have lived five or six million years ago — this means that the neural machinery underlying depression (or grief) could be at least that old! I couldn't help but think, what else could animals tell us about psychiatry?

Ethology is the formal study of animal behavior, but it also sheds light on humankind. It's like trying to deduce the operations of a car engine: it's much easier if you study cars from the 1950s than if you look at complex modern engines. Similarly, animal behaviors can elucidate the basic framework on which hominoid intelligence is built. Even twenty years before publishing *On the Origin*, Darwin appreciated that the study of animals could help illuminate the basic operations underlying humanity. In 1838, Darwin jotted in a notebook, "He who understands baboons would do more towards metaphysics than Locke."

The aspect of animal intelligence most relevant to psychiatry is social acumen. As *Homo sapiens*'s nearest living relative, chimpanzees reveal

---

[9]Negative assortative mating takes place when individuals choose dissimilar mates and reduce variation within the species.

a host of behaviors, both in captivity and in the wild, that appear to be rudimentary forms of contemporary human sociality: sharing, anger, retaliation, rank and leadership, teasing, reconciliation, attachment, communication, alliance formation between non-kin males, tool use, warring and deception (de Waal 2001). Cetaceans (dolphins and whales) are arguably as intelligent as chimpanzees, however their intelligence originates from a slightly different evolutionary lineage. The last common ancestor of primates and dolphins roamed the planet about ninety million years ago (Bearzi and Standford 2008). Dolphins are well known for their sociality and ability to cooperate, exhibiting behaviors which have previously been considered the sole domain of primates: self-recognition, imitation, complex communication, deception and possibly even tool use. (There are instances of female dolphins wearing marine sponges, purportedly to protect against abrasions while scouring the ocean bottom.) Two or three unrelated male dolphins will often form permanent alliances, a behavior otherwise only observed in humans. These male alliances can ward off predators like sharks, as well as help corral females in order to mate (Mann et al. 2000). Disputes between warring dolphin alliances where smaller groups may have to retreat from a potential food source have also been observed (Smolker 2001). The remarkable semblance between dolphins and primates implies a certain degree of inevitability related to the evolutionary progression of social behaviors.

Parrots are another species that show incredible social intelligence. My kids once took in a wayward cockatiel that perversely took an attachment to me, rather than to them. This animal (we named him Nelson) was exceedingly social. He was frequently attached to my shoulder, but also involved himself in every family fray. As Nelson's sweetheart, I was the only family member associated with an easily identifiable signature whistle, "ee-ooh." On one occasion, Nelson appeared to be playing peek-a-boo with me, repeatedly calling me and then hiding behind a picture frame when I came looking for him. I later learned that, according to experimental studies, parrots possess an understanding of object permanence rivaled only by the higher primates (Pepperberg 1999). In other words, I wasn't imagining Nelson's playful behavior. Parrots are extremely social creatures, which means they are constantly monitoring interactions between themselves and others. It is therefore not too surprising that they can play peek-a-boo. What may be even more fascinating is that the last common ancestor between birds and mammals existed about three hundred million years ago. The seeds of sociality have truly been around a very long time!

Other animals such as such as elephants, lions, wolves, geese and hyenas also have interesting social lives filled with various degrees of communication, hierarchies, altruism and cooperation between non-kin

(Wilson 1975). It is abundantly clear that such sophisticated social behaviors are found throughout the animal kingdom, and are not the sole domain of human beings.

Most social behaviors are reflexive and reflect genetically rooted behavioral algorithms, analogous to computer programs. For those of us who live in contemporary Western nations, our fondness for sweet fatty foods can be a burden. However, these powerful cravings would have been entirely appropriate in traditional societies, where starvation rather than obesity was the more pressing problem. This straightforward example can be extended to more subtle social behaviors. Whether or not I decide to loan my favorite book to my boss, for example, will be partially dictated by ancient behavioral programs that consider my tendency to be altruistic, my hierarchical position relative to my boss, and my possessive feelings toward the book. The resulting decision may not be entirely rational.

There are scientific fields wholly dedicated to exploring the evolutionary roots of human social behaviors. Sociobiology, for example, is the evolutionary study of animal behavior with the ultimate intention of explaining human behavior (Wilson 1975). Evolutionary psychology can be viewed as the reverse of sociobiology: it primarily tries to explain contemporary human behaviors through the knowledge gained from ethology, anthropology and evolutionary principles. Even economists cannot afford to ignore ethology and evolution. Thorstein Veblen was one of the first economists to incorporate ethology, anthropology and evolutionary principles into his work. In his classic book, *The Theory of the Leisure Class*, published in 1899, he introduced the now popular concept of "conspicuous consumption," which explained irrational consumer spending in the context of social status signaling. He believed this behavior is rooted in our evolutionary history. Behavioral economist Dan Ariely's recent book *Predictably Irrational* (2008) describes a number of very interesting experimental studies highlighting foolish consumer choices — irrational decisions that are presumed to be rooted in primal instincts.

Love is perhaps our most enigmatic emotion; however, it can be in some measure understood through ethology-based attachment theories. The more unsavory side of love, infidelity, also has some compelling precedents in nature. Across primate species larger testicles tend to correlate with polygamous behavior. On one end of the primate spectrum are male gorillas with their comparatively small testicles, presumably due to the fact that female gorillas are monogamous. In contrast, chimpanzees, well-known to habitually share partners, have large testes proportionate to body size. This makes evolutionary sense because sperm competition is of greater importance to male chimpanzees. What about the size of human testicles? Apparently, they fall somewhere between gorillas and chimpanzees. This seems to indicate that as a species we are inclined to be somewhat monogamous, but not completely — a scenario which is reflected in human behavior across almost every culture (Barash and Lipton 2001).[10] (I should explain that this does not necessarily justify infidelity. Knowing the origins of a behavior can also be helpful in devising the necessary steps to curtail it.)

The origins of many human activities can be traced back to similar animal behavior. The next question is whether animals also share our propensity for psychiatric disorders. In fact, there are a few psychiatric disturbances that seem to have parallels in other animals. The pathological effects of social isolation, for example, are devastating both to humans and to social animals.

In the 1950s and 60s, Harry Harlow conducted a number of notorious primate experiments, typically on rhesus macaques, that were both brutal and arguably invaluable (Blum 2002). In one series of experiments, infant monkeys were raised in complete social isolation, which inevitably induced extreme behavioral disturbances varying from excessive apathy to fervid anxiety states. Other problems included repetitive rocking, self-biting, and profound social incompetence with peers.

Parrots are another social species vulnerable to the devastating consequences of social isolation. Isolated parrots behave erratically, exhibiting self-mutilating behaviors like feather-picking.

In a similar vein, the group of psychiatric patients usually indentified as having borderline personality disorder are well known for their propensity towards self-mutilation and suicidal ideation. Epidemiological studies have shown that almost all of these patients have histories of childhood neglect or abuse (Bandelow et al. 2005; Zanarini et al. 1997). Their self-destructive thoughts tend to occur during periods of social stress, particularly after rejection from an intimate partner. Borderline patients often feel confused by these distressing thoughts and impulses. Psychodynamic explanations could be part of the story, but it is equally possible that primitive neural centers are being reflexively activated.[11]

Clinical depression seems to be associated with a shortlist of possible causes. Three purported etiological factors have compelling parallels in the animal kingdom: learned helplessness, severed social attachments and loss of status.

In the 1960s, Martin Seligman and others conducted a series of experiments showing that placing dogs in situations where they are helpless — for example, applying electric shocks from which there was no escape — would reliably generate depressive-like behaviors. They reasoned that

---

[10] These are inter-species correlations. Testicular size between males of the same species does not necessarily correlate with infidelity.

[11] I surmise that extreme childhood neglect is rare in traditional societies and accordingly, borderline personality disorder may be an artifact of Western societies.

"learned helplessness," when patients perceive that there is no solution to their distress, could explain those types of depression. This dynamic is sometimes seen in depressed people unhappy about their marriages but reluctant to leave their spouse for the sake of the children or because of financial considerations.

Severing a productive social attachment is never evolutionarily beneficial. Juveniles, particularly infant social mammals, become very distressed when separated from their mothers. Youngsters initially express agitation, an emotional state that prompts evolutionary adaptive behaviors such as frantically searching for mother. If a long enough time elapses, depressive-like behaviors follow, perhaps to limit ineffective use of energy or minimize exposure to environmental hazards. Maternal separation anxiety is evolutionarily very old and could have later been co-opted in the service of preserving sexual relationships or reinforcing anti-predatory behaviors. In humans, the sadness of a break-up, for example, may be our body's way of motivating us back into a potentially promising relationship. However, this evolutionary emotional response is only a rough approximation of what previously worked best for our ancestors in similar situations. Other competing feelings or rational predictions will certainly affect the final decision.

Loss of status, in all its various forms — job loss, financial ruin, failed exams — is another common reason for depression. According to John Price, one of the pioneers of evolutionary psychiatry, many depressive responses could originate in animal appeasement displays. Price and his co-authors describe how animals often demonstrate submissive behaviors when they figure that a dispute has become a lost cause. Adult wolves may act like puppies to submit to a senior wolf in the pack hierarchy; male primates may act feminine and offer their genitals as they submit to a superior male, who may even display a token mounting action. Even reptiles have been known to make appeasement displays, which implies that submissive behaviors have been around for several hundred million years. In humans, certain forms of depression could signal that one is retreating from competition and is therefore no longer a threat (Price et al. 2004).

A glimpse of the evolutionary trajectory of obsessive-compulsive disorder (OCD) might be seen in cats, dogs, horses and parrots (Polimeni et al. 2005). For example, canine acral lick dermatitis is a condition where dogs exhibit repetitive self-licking and chewing to the point of self-mutilation. It most often emerges in the context of stressful captive environments and is probably rare in the wild. Interestingly, it improves with serotonergic antidepressants, the same treatment as for human OCD.

Do animal models exist for schizophrenia? I am doubtful that the core features of schizophrenia are even vaguely represented in animals. However, it is known that the administration of amphetamines can exacerbate psychotic symptoms in schizophrenia and that such psychostimulants can also produce hyperactive behaviors and stereotypic movements in rats. Moreover, stereotypic movements can be attenuated by antipsychotic medications. So there is certainly a similarity in one aspect of schizophrenia, but we are a long way from a comprehensive animal model. (Marcotte et al. 2001; Lipska 2004; Lipska and Weinberger 2000).

Evolutionary Psychiatry

Ethology led to the field of evolutionary psychology, which in turn spun off evolutionary psychiatry. One of the most useful ideas associated with evolutionary psychology is the concept of cognitive modules. During the process of brain evolution, higher cortical structures could have, in theory, recruited older neural complexes to solve novel evolutionary problems (Geary 2004) — the *modularity* refers to neurobehavioral functions, not to specific anatomical locations. An analogy is found in computer programs: the purpose of an Internet browser is to access the Internet, but to do so effectively, the program needs to incorporate a multitude of smaller programs such as video convertors and text processors. In a similar vein, it has been proposed that jealousy is seated in a cognitive module activated when a sexual mate flirts with another potential suitor. The hypothesized jealousy module may be connected with other, more rudimentary modules such as the attachment or anger centers. Cognitive modules could support such survival functions as sexual arousal, mate choice, the fight-or-flight response, language and even shamanistic (i.e., schizophrenic) behaviors.

I have already mentioned that Julian Huxley and Ernst Mayr (two non-psychiatrists) were the primary authors of the first evolutionary psychiatry paper, in 1964. Although they got the ball rolling, the next three decades saw relatively little work in the field. However, by the early 1990s, about a dozen psychiatrists began to look at psychiatric ailments through the prism of evolutionary psychology. Randolph Nesse and George C. Williams's classic popular science book *Why We get Sick*, published in 1994, contained a brief chapter about mental disorders. Anthony Stevens and John Price published the first dedicated textbook entitled *Evolutionary Psychiatry* in 1996, followed by Michael McGuire and Alfonso Troisi's *Darwinian Psychiatry* in 1998. Martin Brüne's *Textbook of Evolutionary Psychiatry*, published in 2008, is the most recent general textbook and an excellent resource for those new to the field. These pioneering evolutionary psychiatrists established the basic scientific parameters that explain how evolutionary theory may integrate with mental disorders. Presently, there are several evolutionary hypotheses for each major psychiatric diagnosis including depression, schizophrenia, obsessive-compulsive disorder, bulimia, personality disorders and PTSD. As in all pioneering fields, many of these theories are speculative and undoubtedly off the mark. However, as the field matures, some of the better theories will eventually gain prominence and become integrated with the rest of mainstream psychiatry.

The psychiatric diagnosis most intensively studied by evolutionary theorists has been schizophrenia. Over the last few decades, a dozen or so researchers have speculated on the possible evolutionary origins of schizophrenia. I will sidestep a review of these ideas because they tend to be varied and intricate, and will take us way beyond the scope of a book about shamanism. (See the following reviews for peer-reviewed critiques of evolutionary ideas pertaining to schizophrenia: Polimeni and Reiss 2003; Brüne 2004; Pearlson and Folley 2008.) I will, however, review one alternative proposition because it clearly crosses paths with the shamanism theory.

Stevens and Price (2000) developed a group-splitting hypothesis of schizophrenia to explain how human tribes disperse into new environmental niches. Dispersal of a species is a very important function of evolution. For hominoid species, this means that tribes must eventually split (Price 2009). According to Stevens and Price (2000, 6):

> ...the schizotypal genotype is an adaptation whose function is to facilitate group splitting, and that this is achieved by the formation of a subgroup or cult under the influence of a charismatic leader, who is enabled by his borderline psychotic thinking to separate himself from the dogma and ideals of the main group, and to persuade his followers that he or she is uniquely qualified to lead them to salvation in a "promised land."

Stevens and Price found schizotypal personality traits in such cult-like leaders as Reverend Jim Jones of Guyana, David Koresh of Waco, Texas and even Adolf Hitler.

When Jeff Reiss and I published our shamanistic theory, we were unaware that John Price and Anthony Stevens had, two years earlier, formulated their group-splitting theory of schizophrenia. (Their theory was not published in a peer-reviewed journal.) Although each theory drew a different conclusion, the supporting arguments were intriguingly similar. It was John Price who figured out how to reconcile the theories, suggesting that shamans and prophets could be two phenotypic versions of a schizotypal personality. In an email exchange (April 30, 2011), John Price wrote:

> Can we say that in the phase of group expansion the magico-religious practitioner (schizotypal personality) becomes a shaman

and promotes group cohesion, but when the group gets too big he or she becomes a prophet and encourages splitting? For group splitting at the hands of the cult leader, we envisioned a vector leading from the original group out towards a promised land — resulting in separation of the daughter group from the old group. In-group paranoid delusions would contribute to dispersal, and there were also delusions of something better elsewhere — ideas of "making the desert bloom" (one of my patients), of a "land without evil" across the mountains (common in Brazilian prophets) and other descriptions of a "promised land" (e.g., Abraham). Among the prophets there was also a cognition of the existing society being evil in some way (not necessarily delusional), and therefore something to be got out of or away from. Out-group paranoid delusions fit the shaman better than the prophet, because the shaman is encouraging group cohesion, the better to fight the out-groups (and so make more room for future daughter groups?).

I find this synthesis wholly reasonable, although my own view is that there is more evidence pointing towards analogous behaviors between shamans and schizophrenia. This may be due to the fact that, in hunting and gathering societies, schizotypal personalities would have inevitably spent a lot more time as shamans than prophets. Group splitting is a relatively uncommon event in the life of a tribe (although it may not be any less evolutionarily important). The possibility of such a dual hypothesis is intriguing and could perhaps lead us to an entirely new set of research questions. For example, how do schizotypal personalities switch from shamans to prophets? Would it depend on rational social circumstances, or could there be underlying genotypic triggers that prompt the development of one type of delusional over another?

Anthropology

The study of hunting and gathering societies provides a window into ancestral hominid environments and the selection pressures that shaped our various cognitive skills. A teenager listening to music on an iPod offers no clue as to the possible evolutionary purpose of music. In contrast, a feverish tribal dance in anticipation of battle offers a hint. Before we consider shamanism, it is necessary to consider four anthropological topics: human universals, shamanistic art, tribal warfare and hominid social traits seemingly linked to group selection.

Human Universals

A number of human behaviors are seen in every culture, including traditional societies. These "human universals" are of particular interest because they may reflect genotypic adaptations (Brown 1991). For example, the fight-or-flight response prepares organisms for impending physical exertion by initiating physiological changes, including increased heart and respiratory rate. It is a biological response that is universal, reflexive (instinctive) and obviously evolutionarily adaptive. This fight-or-flight reflex is the product of dedicated neurophysiological pathways, ultimately controlled by a genetic substrate amenable to the forces of evolution.

Of course, not all universal behaviors are the culmination of genetic change. The use of fire, for example, is so undeniably useful that no neurocognitive reflex or genetic blueprint is necessary to maintain its practice.

Certain complex human traits, such as shame, envy or sexual jealousy, are also universal and seem to be invoked instinctively. In other words, we do not feel shame because we have made a logical decision to do so. It simply happens, reflexively, in certain social situations. The instinctual component tends to be overlooked because shame is a complex emotion laden with a lot of "post-game analysis." If shame is truly reflexive, it likely involves dedicated neural pathways created by evolution.

In contrast, some human universals may be epiphenomena with no evolutionary purpose. For example, attempting to supernaturally control the weather is apparently a behavior found in all cultures. However, the human brain's tendency to believe in supernatural phenomena exists for reasons entirely separate from concerns about the weather. Once people began thinking in supernatural terms, applying such thoughts to something as common and unpredictable as the weather seems inevitable. Thus, this is likely an incidental behavior, with no appreciable positive or negative selection pressures associated with it. It is doubtful that we possess "try to control the weather" genes.

Shamanism, however, is not likely to be an incidental behavior. Instead, shamanistic behaviors seem to have served specific survival functions in ancient tribes. Psychosis will often (reflexively) overcome certain people who eventually become designated as shamans. It is also wholly possible that there are specific neural pathways (associated with dedicated genes) that sustain the entire shamanistic process.

Human universals are important because very little *a priori* knowledge is required to make a strong case that a universal trait is evolutionarily adaptive. Let's assume that we know nothing about shamanism and instead simply refer to it as behavior X. If we can agree that behavior X is 1) universal (occurring throughout the world over many generations) and 2) instinctive (not the result of a learned behavior), this means that the function is almost certainly evolutionarily adaptive. Notice that we have put this together without knowing anything about *how* shamanism could be adaptive.

Shamanistic Art

Archeologists study ancient artifacts to speculate about the lives of our ancestors (Tattersall 1995). Archaeological remains can theoretically help trace the progression of higher order thinking — including shamanism. For example, the quality of tool technologies has inevitably progressed through hominid evolution; Mousterian tools (300,000 years old) are superior to Acheulean tools (one and a half million years old), which are finer than Oldowan tools (two million years old). Tools, however, do not provide much insight into broader thought processes. The advent of language, for example, cannot be easily ascertained by scrutinizing prehistoric hand axes.

Prehistoric art, on the other hand, does provide clues about the progression of higher cognition in hominid history. The first evidence of art — clusters of cupules (man-made pits on rock surfaces) or poorly discernable statuettes resembling human figures — arrives (debatably) on the archeological scene between 50,000 and 100,000 years ago. Because of their simplicity, not all archeologists are convinced that these artifacts can truly be considered art. However, beginning about 40,000 years ago, a distinct change appears in the archeological record. Elaborate artworks suddenly flourish among ancient relics — implying a new level of symbolic thought (Mithen 1996). One of the oldest and most striking examples is an elaborate statuette of a lion's head merged with a human lower body which was found in Germany (Figure 6.2). It was carved out of mammoth ivory and is estimated to be about 32,000 years old. The Chauvet cave paintings located in south-central France are also believed to be about 32,000 years old (Klein and Edgar 2002). These caves, discovered in 1994, contain hundreds of detailed paintings, many of which demonstrate subtle shading and hints of perspective. Cognitive archeologist Steven Mithen (1996) surmises that such art represents the first evidence of anthropomorphic as well as totemic thinking. At the very least, it seems to represent some sort of evolutionary development in the fluidity of thought. (The term "cognitive fluidity" is, of course, vague; the precise psychological changes that ushered in the ability to create art are not yet known.)

There are very specific reasons why some anthropological experts believe shamans created much of this prehistoric art (Dowson and Porr 2001; Lewis-Williams 1997; Pearson 2002; Yates and Manhire 1991). First,

Figure 6.2: The Lion-man of Hohlenstein Stadel, Germany is believed to be 32,000 years old (Wikimedia Commons September 1, 2011).

cave paintings bear a strong resemblance to contemporary aboriginal art, which is usually created by shamans. Fusing animals with human beings is a common contemporary shamanistic theme which is also frequently found in prehistoric art. Other shared characteristics are elongated or grossly distorted body parts and collage-style scenes. It has even been claimed that the Fumane Cave near Verona, Italy may contain a 35,000-year-old image of a shaman (Rossano 2010; Broglio et al. 2009), but to my untrained eye, the depiction seems indistinct and unconvincing.

Entoptic images are specific visual designs derived from hallucinatory experiences. Experts of prehistoric art have noticed distinct entoptic images running through both modern shamanistic art and archeological finds (Dowson and Holliday 1989; Lewis-Williams 1997; Stahl 1986). (As we shall see in Chapters 8 and 9, the use of hallucinogenic substances is extremely common in contemporary shamans.) Drug-induced hallucinations often generate specific diagrammatic patterns, probably due to the physiological constraints of the visual cortex (Bressloff et al. 2002). Although each expert classifies them slightly differently, certain shapes are common: zigzags, multiple circles, spirals, wavy parallel lines and lattice designs (Devereux 1997; Dowson and Holliday 1989). In contrast, squares or rectangles are rarely seen among entoptic hallucinations, and accordingly such shapes are rarely observed in prehistoric or modern shamanistic art. Squares and rectangles, however, are regularly found in secular modern art.

Another compelling feature of prehistoric art is that it appears to only deal with a handful of themes. A survey of the several hundred painted figures found in Europe's three most famous prehistoric caves — Altamira in Spain, and Lascaux and Chauvet in France — yields a short list of motifs. Most paintings concern animals of the hunt, such as bison, deer and horses (Figure 6.3). Fierce animals like panthers and bears were also drawn. This theme is remarkably similar to contemporary shamanistic art, which commonly involves prey animals or dangerous predators like panthers, lions or eagles. It is interesting that illustrations of plants and landscapes were almost non-existent, and drawings of humans sporadic. The conspicuous absence of mundane subjects in ancient artwork hints at spiritual concerns. Moreover, many ancient carvings are fragile, which implies ritualistic rather than domestic use (Ryan 1999).

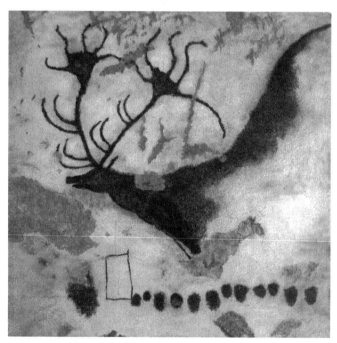

Figure 6.3: This painting of a deer from the Lascaux caves is about 17,000 years old (Wikimedia Commons September 1, 2011).

The advent of art about 40,000 years ago approximately coincides with the modern human migration out of Africa 50,000–60,000 years ago. This new kind of hominoid was able to push Neanderthals to extinction (Kingdon 1993). I cannot imagine that it was art that defeated the Neanderthals, but perhaps something else related to this new fluidity of thought. This ineffable cognitive attribute — fluid thinking — is perhaps also seen in religion, dance, music, rhyming, and humor. Each of these behaviors conspicuously enhances group cohesion, and such tribal solidarity would have been particularly useful in war.

Tribal Warfare

In this section, I put forward the thesis that inter-tribal war — over a period of at least several hundred thousands of years — generated immense selection pressures, leading to remarkable changes in hominid behavior. Social behaviors such as dance, music, religion and shamanism enhanced group cohesion and ultimately helped tribes wage war against each other. It would appear that all of these traits are, at least partially, supported by group selection.

Zoologists believe that the primary reason animals form herds is for protection from predators (Hamilton 1971; Miller 2002, 96). The protective features of herding may be having more members to watch for threats or to mob attacking predators. Accordingly, the largest aggregations of animals are seen among those that are regularly preyed upon, such as bison, zebras, horses, geese, antelopes, and some types of fish. But as herd numbers increase, the group must travel greater distances in order to feed everyone (Wrangham and Peterson 1996; Janson and Goldsmith 1995). These opposing factors counteract to produce an ideal herd size for any given species.

Predatory animals, in contrast, tend to be solitary or prowl around in small groups. A pride of lions generally hunts in groups of three or four (Stander 1992). Wolves are among the most social predators and their packs rarely exceed a dozen individuals. The spotted hyena congregates in one of the largest predatory group sizes among mammals: their clans sometimes have thirty to forty members. The reason for such tight limits on predatory group size is that the weakest members of the group usually receive the last morsels of the kill, so they are usually better off abandoning their associates and going it alone.

All primates must worry about being eaten, especially when they are alone. Lower primates, in particular, are better characterized as species of prey (although they occasionally display predatory behavior). Their communities are generally comprised of ten to forty individuals (Chapman et al. 1999). Leopards, crocodiles, eagles, snakes and larger primates are their most common predators.

Hominid evolution is accompanied by a transition from being a prey species to being one of a more predatory disposition, although the experts do not agree precisely when our ancestors graduated from scavenger to hunter. Hunting seems to have been a meaningful endeavor for at least a hundred thousand years and probably longer. This transition from relative prey to relative predator was not, however, accompanied by a reduction in group size. In fact, it appears that just the opposite occurred. Contemporary human tribes most often have a population of thirty to one hundred fifty individuals, and some groups approaching three hundred people have been reported by anthropologists. This is uncharacteristic for a predatory species.

So, who could have been the primary predator of ancient hominids? Although lions and tigers were always a threat, hominids have had spears for at least 400,000 years (Otterbein 2004, 50). In my view, there wouldn't be any advantage to mobbing a predator with a group larger than the number of people that can physically encircle an animal: not much more than a half-dozen adults. So why this extra herding? A number of researchers believe that humankind's greatest predator was man himself — specifically, men from other tribes (Alexander 1987; Bowles 2009; Choi and Bowles 2007).[12]

There is overwhelming evidence that *Homo sapiens* have had a very violent past. Numerous anthropological studies have shown that almost all hunting and gathering societies (Otterbein 2004, 82) engaged in regular warfare with other tribes. Only a handful of relatively isolated tribes like the Copper Inuit of northern Canada or the Dorobo peoples of Africa are known to have lacked formal military organization, although they would fight to protect themselves. Where reliable estimates could be obtained, traditional societies often saw a death rate from war of 0.5–1.0 percent of the population per annum. This translates to 10–30 percent of total male deaths resulting from warfare (Keeley 1996, 90).

Military historians, like Martin Van Crevald (1991), have distinguished two modes of war: ritualistic war (perhaps to test the strength of respective armies) and raids, usually accompanied by much higher death rates (Van Crevald 1991, 74). Surprise attacks (raids, ambushes) were probably the most effective method of warfare. For example, according to Keeley, "nearly all western North American Indian groups were raided at least twice each year" (Keeley 1996, 66). Women and children were not necessarily exempt as targets of raids.

The archeological record shows signs of significant violence in ancient prehistoric peoples (Ehrenreich 1997). In thousand-year-old prehistoric sites in central California, Keeley noted that 5 percent of skeletons contain embedded arrows. Because not all war deaths result in a projectile clearly

---

[12]Even Darwin believed that having talented warriors could help tribes "supplant other tribes" (Darwin 1871).

embedded into the bone, the actual rates of death-by-arrow may have been as high as 40 percent (Keeley 1996, 91). Other researchers have projected similar frequencies, such as a 27 percent violent-death rate in 1500-year-old skeletons from the British Columbia coastline (Maschner 1997, 273).

One military weapon, invented about 40,000 years ago, may have considerably changed the course of inter-tribal fighting, especially between Neanderthals and *Homo sapiens*. The atlatl is a device that, through leverage and spring action, dramatically increases the speed and accuracy of spear throwing. Equally effective against animal or hominid foe, the atlatl made killing easier. The archeological record suggests that Neanderthals did not possess this weapon, which could have only meant they would have been awfully disadvantaged in battle.

Even the domestication of the dog may have been motivated by war. Wolves are believed to have been domesticated about 14,000 years ago by hunting and gathering societies of the Middle East or perhaps central Europe. Although dogs can be helpful during hunting expeditions, their role does not appear to be crucial. However, dogs are unique in their predisposition to vocalize loudly at the first sign of an approaching unfamiliar creature. According to Driscoll and Macdonald (2009), "Thereafter, these wolves may have found utility as barking sentinels, warning of human and animal invaders approaching at night."

All this warfare must have created immense selection pressures — an evolutionary drive to advance phenotypic traits whose possessors could not only defend against the onslaught of an invading tribe but also disperse the species through effective attacks. One such adaptation could have been the "unusually thick skulls of *H. erectus*" (Striedter 2005, 318). Another obvious solution is larger tribes; however, this causes its own problems: it jeopardizes group solidarity and hinders food procurement.

Sustaining group unity would be vital in a milieu of feuding tribes (Puurtinen and Mappes 2009). It turns out that enhancing group cohesion is like an evolutionary two-for-one sale. A cognitive attribute such as language not only facilitates larger numbers per tribe but it also turns any size of group into more effective warriors. Greater tribal solidarity likely helped the physically smallish *Homo sapiens* displace the larger and stronger Neanderthals.

Human Social Traits Likely Supported by Group Selection

There are a number of social traits that seem to be both instinctual and well-suited to creating group unity. Some of these group-selected phenotypic traits are language, humor, music, dance, in-group versus out-group bias and religion. I will explore religion separately in the next chapter.

Language

Although language is undoubtedly useful for any species, an obvious question is, why has it only evolved once? Robin Dunbar argues that language would have been particularly useful for larger primate groups to preserve the emotional bonds that were no longer maintainable through grooming behaviors associated with smaller primate communities (1993). Perhaps the development of better tools over the last two million years made humans more efficient (and deadly) competitors, resulting in an evolutionary pressure towards larger groups. Language, therefore, could be the evolutionary solution to the problem of maintaining the social cohesion and therefore the existence of large predatory groups.

Humor

In a paper entitled, "The First Joke: exploring the evolutionary origins of humor" (2006), Jeff Reiss and I estimated humor to have been around for at least 40,000 years, since we know humor is present in Australian Aboriginals whose genetic lines have been separated from the rest of humanity for 40,000 years.

Humor is both universal and invoked reflexively. Like Chomsky's universal grammar, humor cannot be learned without some pre-wired cognitive circuit. According to a reputable humor theory by Thomas Veatch, all humor involves themes of social violation. Humor, it seems, injects positive feelings while hierarchal competition and other minor social violations are being worked out — pretty powerful stuff to boost group cohesion (Weisfeld 1993).

Music

Every fall, several hundred Canada geese settle down in the evenings near my home on the Assiniboine River in the Canadian prairie. Each morning, a few geese begin to honk. As others join in, the geese become increasingly excited and the noise rises to a cacophonous crescendo, which at its peak culminates in a throng of synchronized departures. It is obvious that the escalating honking coordinates an excitatory state that prepares the geese to ascend as a unit. Synchronized chorusing has been identified throughout nature, including in croaking frogs and various insects (Merker 2000).

Similarly, music allows large groups of people to coordinate their emotional states. Major chords evoke happiness across all cultures, while minor chords invariably provoke sadness. Fourth (IV) chords create an ineffable tension that only resolves by proceeding to the major (I) chord (for example, in the verse of the song "Pinball Wizard" by The Who, or the "amen" cadence at the end of many hymns). Audiences at concerts run through coordinated emotional states dictated by the musical performance — sadness during a poignant song or happiness during a raucous grand finale. As music scientist Daniel Levitin (2008, 146) writes, "music is probably the most reliable (non-pharmaceutical) agent we have for mood induction." Anthropologists have frequently commented on how, in traditional societies, music is predominantly associated with spiritual ceremonies and battle preparations (war dances) (p. 194) — situations when the emotional coordination of a tribe would be paramount.

Dance

Dance often accompanies music, especially in traditional societies. Moreover, prehistoric art often depicts dancing in similar circumstances as it is seen in contemporary traditional societies (Garfinkel 2003). As through music, it seems that unspecific excitatory moods can be easily conveyed through dance. Repetitive rhythmic dancing is often associated with shamanism and trance induction. Dancing is typically woven into spiritual ceremonies, shamanistic trances and war preparations.

In-group versus out-group bias

Social identity theory is a branch of psychology that investigates our reflexive inclination to side with our own groups and alienate others (Brewer 1999; Efferson et al. 2008). Religious institutions, especially smaller ones, are notorious for their in-group loyalties and out-group wariness (as are tribes, armies, threatened ethnic groups and sports teams). There is overwhelming evidence that killing at a greater physical or psychological distance — thereby putting the victim in the out-group — is substantially easier for all but a tiny minority of human beings (Grossman 1995).[13]

Superimposed on tribalism (or perhaps a separate evolutionary mechanism) are the positive feelings that make battles exciting and alluring. General Robert E. Lee once said, "It is well that war is so terrible — otherwise, we would grow too fond of it!" Three distinct positive emotions seem to be especially associated with the act of war: excitement (the adrenaline rush of battle), camaraderie and glory. These agreeable feelings are also the cornerstones of team sports. If one were to observe team sports with a dispassionate eye, one might wonder why anyone would bother with these spectacles. It is our instinctual predilection to side with our own

---

[13]Some authors have talked about morality as a device for group cohesion. I have purposefully sidestepped this term because it is so imprecise — it's almost as ineffectual as describing someone as "good." I prefer to discuss the conjectured elemental components of "moral" behaviors, such as altruism, empathy, conformity, shame, and other rudimentary feelings.

group and battle with others (with all the associated emotional highs and lows) that perhaps best explains every modern society's passion for team sports. Many team sports are structured to represent war, and discussed using terms like victory, defeat, defense, counter-attack, run-and-gun offence, throwing the bomb, and perhaps most to the point, "sudden death" overtime. It is as if evolution installed a cognitive program for us to feel good about engaging in team-fighting behaviors. This kind of pre-wired evolutionary encouragement would have been immensely helpful when standing your ground in the face of the prospect of getting your head knocked in.

Conclusion

This chapter surveyed basic evolutionary theory. The science of phenotypic variation received special attention due to its potential relevance to psychiatry, the field of medicine that focuses on deviant social behaviors. We also singled out group-selected task specialization because of its presumed role in shamanism. This was followed by a brief review of the fundamental teachings of ethology, evolutionary psychology and evolutionary psychiatry. Last, we explored the daily lives of our ancient ancestors in order to better understand the evolutionary pressures behind higher order thinking. Language, humor, dance, music, tribalism, shamanism and religion could very well have been group-selected traits, ultimately used in the service of inter-tribal competition.

– 7 –
The Evolutionary Origins of Religion

Religion is a part of every society and accordingly thousands of religions have existed throughout history, from familiar monotheistic churches to scores of new religious movements, as well as the distinctive spiritual beliefs associated with even the most remote hunting and gathering societies. Perhaps the most succinct definition of religion has been supplied by Pascal Boyer: "Religion is about the existence and causal powers of non-observable entities and agencies" (Boyer 2001, 7). Agency refers to the "abstract quality that is present in animals, persons and anything that appears to move on its own accord, in pursuance of its own goals" (Boyer 2001, 144). No matter how different two religions may seem on the surface, Boyer's general definition captures a common thread.

An Anthropological Examination of Religion

The diversity of religion is analogous to the wide range of games seen throughout the world. For example, the relationship between Catholicism, Hare Krishna, and Haitian Vodou can be compared to soccer, baseball and kabaddi (an Asian team game with elements of wrestling, chasing and tackling, but no ball). Each sport appears wildly different from the others, but they are linked by a common thread: a desire to compete with others.

Religion and sport share deeper similarities. Both activities begin with subtle internal feelings (or instincts) among certain individuals that coalesce into group activities or formal institutions. Every civilization has witnessed the competitive cravings of its most ambitious members turn into communal games. Similarly, every culture has seen the private spiritual feelings of a few develop into official religious ceremonies.

It is remarkable how a handful of instinctual behaviors reliably develop into group activities in traditional societies. In addition to religion and sport, our innate musicality is an example of this (Mithen 2005). Music seems evolutionarily designed to be contagious, reliably leading to tribal singing and dancing. In contrast, there is no evidence that other instinctual behaviors like gossip, humor, and sex lead to formal institutions in traditional societies. (In larger complex societies they sometimes do. For example, in Western societies, gossip is institutionalized in the form of tabloid magazines or entertainment television news. Humor is, in some measure, socially formalized through comedy clubs and television sitcoms. Sex can be considered ritualized in brothels and sexual fetish clubs.)

The progression from shared feelings to official group activity could be nudged along by evolutionary forces. Throughout the animal kingdom, mimicking is a pervasive behavior which seems to be "turned on" during evolutionarily desirable situations. Hearing musicians jam or watching a pickup game of basketball will prompt many of us to join in. In contrast, seeing my neighbor put up her Christmas lights doesn't get me pumped up to do the same.

Our inclination to participate in sports could be a vestigial behavior which helped our ancient ancestors survive (Lombardo 2012). Tribal games tend to be taken quite seriously and often tactically resemble vital activities like war and hunting. For example, rarajípari, a kickball game played by the Tarahumara people of northern Mexico, requires the running of very long distances, sometimes in excess of marathon distances, by teams of three to forty players. Elaborate rituals, wagering and drinking *tesgüino* (Tarahumaran beer) were part of the spectacle. Kendall Blanchard has suggested that the game of rarajípari could have helped prepare for hunting because, "[t]raditionally, the Tarahumara hunted deer by pursuing them on foot, relentlessly chasing a single animal until it fell over in exhaustion" (Blanchard 1995, 136).

Similar examples can be found in other traditional societies. The Cherokee used to play a game similar to lacrosse that "...involved elaborate preparations by the entire community, fasting, religious ceremonies and celebrations, incantations and magical manipulations, and heavy wagering on the outcome of the games" (Blanchard 2000, 145). Team sports like lacrosse, soccer, basketball, hockey and football require the same skills that are necessary for war: agility, strength, stamina, cunning and determination.

In comparison to sports, the evolutionary purpose of religion is more elusive. Although deeply religious people may feel this to be irrelevant, anyone who doubts the existence of God is in a position to consider an evolutionary role for spirituality. In fact, spirituality has several characteristics that suggest it could have been evolutionarily advantageous. Spirituality seems to originate in instinctual (and not necessarily rational) beliefs, and results in social behaviors that could have increased survival rates in traditional societies.

However, the austere churches of modern civilizations do not readily suggest an evolutionary purpose for religion. A number of social benefits, such as relief of emotional suffering through confession, efficient sharing of resources, or establishing guidelines for harmonious living (e.g., the Ten Commandments), could be derived from the major monotheistic religions (Christianity, Judaism and Islam). However, monotheism has only existed for about 3000 years — a relatively small period of time on evolutionary timescales (Ferm 1965). Robert Wright, author of *The Evolution of God* (2009a, 17), claims, "there is no such thing as an indigenously monotheistic hunter-gatherer society." Therefore, in order to properly explore the evolutionary origins of religion, we must study spirituality in its natural state: the polytheistic religions of hunting and gathering societies.

Anthropologists have documented hundreds of the religious belief systems of hunting and gathering societies. This is a very large sample size for gleaning universal themes. The polytheistic religions of traditional societies undoubtedly share certain similarities to contemporary monotheism: both deal with supernatural entities and include rituals concerned with food procurement and life transitions such as birth, marriage and death. However, notable differences also exist. In contrast to the major monotheistic religions, the religions of traditional societies frequently have the following characteristics:

- multiple gods
- gods depicted as more human or animal-like
- gods more integrally involved in common daily activities, such as tomorrow's hunting expedition or next week's weather
- magical or supernatural events are more likely to be influenced by ordinary individuals (including shamans) as opposed to celebrated prophets
- dreams regularly contain premonitory omens
- malevolent spirits are often blamed on unfriendly persons or hostile tribes

In the late nineteenth century, British anthropologist Everard Thurn described the native tribes of the former British Guiana in his book *Among the Indians of Guiana* (1883), a typical anthropological account of a hunting and gathering society. Thurn spent considerable time with the Macushi tribes and what stood out for him was the prominence of spirituality (that is, animism) in every facet of aboriginal life. All of the common terrestrial objects were represented by well-known spirits such as the sun-spirit, moon-spirit and water-spirit, and a large assortment of lesser spirits also existed. In fact, every object, plant, animal or person could be imbued with its own supernatural spirit. Even notable events, such as an inopportune flood, were sometimes ascribed their own spirit.

There were harmless spirits, malevolent spirits and goodwill spirits for protection. The Macushi were particularly apprehensive about malevolent spirits. In fact, so concerned were they that a specific term existed for nasty spirits: *kenaima*. Anything could possess kenaima — humans, animals, plants or objects. Whenever a person was sick, the *peaiman* (shaman) would try to remove a caterpillar or stone from the ailing victim, which supposedly eradicated the germ of the imposing spirit (Thurn 1883, 349). The Macushi were even concerned with mundane objects such as certain darkened glazed rocks on the bottom of shallow riverbeds. Thurn wrote, "Whenever I questioned the Indians about these rocks, I was at once silenced by the assertion that any allusion to their appearance would vex these rocks and cause them to send misfortune."

In all cultures, rituals inevitably follow spiritual beliefs. The Macushi, for example, partook in a ritual that involved rubbing red peppers in every participant's eyes, including children's, in order to ward away evil spirits originating in other tribes.

Aboriginal folklore often contains stories where the distinction between humans and animals is blurred, like that 32,000 year-old lion-person statuette found in southwestern Germany. Thurn recounts a Macushi story about a mythological tribe where the inhabitants were "men by night but fish by day" (Thurn 1883, 384).

Not only do people and animals transform into each other but dreams can also meld into reality. Thurn tells a story about a Macushi man who dreamt that he had been forced to lug a canoe while feeling sick. In fact, the Macushi are excused from such chores when feeling unwell. In the dream, however, it is Thurn who coerces the ailing man to transport the canoe. Upon awakening, the Macushi man, visibly upset, incredulously scolded Thurn for his crass behavior (which only existed in his dream).

These anecdotes from the Macushi tribe are neither special nor unique, but represent ordinary examples of spirituality in hunting and gathering societies. Similar customs were observed on the other side of the globe. James Dawson studied a number of tribes in the Western District of Victoria, Australia during the late nineteenth century. He described a number of spirits such as *Pirnmeheeal*, a good spirit depicted as a gigantic man living in the clouds, and *Muuruup*, a bad spirit who lived underground but frequented bushes and scrubs at night. Muuruup would sometimes send owls to watch people and report back to him. There were also several suspicious terrestrial spirits resembling "devils, wraiths, ghosts and witches" (Dawson 1881, 50). As is so often the case, evil spirits usually originated in neighboring tribes. According to Dawson, "Natural deaths are generally — but not always — attributed to the malevolence and the spells of an enemy belonging to another tribe" (p. 63).

Australian Aboriginal folklore also contains the kinds of animistic beliefs found in other traditional societies (Dawson 1881, 54). Dawson recounts one such legend behind the domestication of fire. According to one of the tribes he encountered, it was the crows that were the first to domesticate fire, using it mostly to amuse themselves. However, a small "fire-tail wren" stole the fire from the crows; then it was stolen again by a hawk who set the countryside ablaze, thus making fire available to man.

Like all traditional societies, Australian tribes performed the typical communal ceremonies surrounding events such as initiation rites, marriages and funerals. Like the Macushi, they believed in the prognostication of dreams (Dawson 1881, 52).

Despite all these superficial differences, the belief systems of traditional societies are fundamentally equivalent to modern monotheistic religions because both are characterized by a belief in supernatural entities. However, there are a few notable differences that distinguish traditional polytheistic religions from modern monotheism. For the purpose of understanding shamanism, two distinctions are worth emphasizing.

First, the religions of hunting and gathering societies tend to be less formalized than modern monotheistic religions. Magic curses, for example, are endlessly generated and tailored to suit particular circumstances. Rituals and folklore are much more susceptible to modification according to the immediate spiritual experiences of each shaman. In contrast, clergymen are simply not allowed to ad lib like shamans.

The second point of difference is that prehistoric religions tend to be concerned with malevolence rather than the omnipotent or benevolent deity more characteristic of monotheism. Although protective spirits are sometimes described in hunting and gathering belief systems, the majority of spiritual concerns involve environmental hazards or the malice of disagreeable people or hostile tribes.

The remainder of this chapter is divided into two sections. The first is about intra-psychic spirituality — the spirituality within an individual's mind. The second concerns the sociological aspects of religiosity. This format is perhaps the most useful way to conceptualize the evolutionary underpinnings of religion. The two ideas — personal spirituality and organized religion — are like itching and scratching. The internal experience of spirituality is like an itch, while the collective social behaviors prompted by the itch — that is, formal religion — are like scratching.

Intra-psychic Spirituality

There are myriad terms referring to various aspects of internal religious experience: transcendental, numinous, mystical and revelatory (Paloutzian and Park 2005). Spirituality is a notoriously vague term because it sometimes refers to personal transcendental feelings and at other times refers to informal beliefs about supernatural phenomena — in other words, both the itch and the scratch.

Although I consider myself an atheist, I remember once having experienced transcendental feelings for a few hours (and therefore became an agnostic for a very brief period). I was sitting at a piano on a lazy weekend afternoon when, in my boredom, my mind began to drift towards existential issues. I recall experiencing a distinct feeling that there must be something greater somewhere out there in the universe. This feeling dissipated after a few hours and thereafter, I never thought much about it (and probably would not have even recalled the incident if it weren't for writing about spiritual matters).

Such ineffable feelings are not rare or unique. But, due to their subtlety, relative infrequency, absence of accompanying physiological parameters and generally indescribable nature, they do not readily lend themselves to be scientifically studied. William James's classic book *The Varieties of Religious Experiences*, published in 1902, was among the first serious academic forays into these sorts of personal spiritual experience (Wade 2009, 8). More recent studies show, rather consistently, that about a third of the general population feels they have had a unique spiritual experience at some juncture in their lives (Spilka et al. 2003). For example, the question, "Would you say that you have ever had a 'religious or mystical experience' — that is, a moment of sudden religious awakening or insight?" has been asked of general populations on several occasions. In a number of U.S. studies conducted throughout the 1960s and 70s, and with sample sizes of between 1200 and 3200, 21 to 53 percent of respondents answered in the affirmative (Spilka et al. 2003, 309). Similar questions from other studies have yielded approximately 35 percent positive response rates (Spilka et al. 2003, 310–11). These figures are even higher in explicitly religious populations (Spilka et al. 2003, 308). For instance, the question, "Have you ever as an adult had the feeling that you were somehow in the presence of God?" was posed to several thousand church members in the San Francisco Bay area in 1963, with 72 percent responding affirmatively.

One of the first experiments attempting to explore mystical experiences in a controlled scientific setting was designed by Walter Pahnke under the academic supervision of Harvard's Timothy Leary (Doblin 1991). In 1962 twenty Protestant divinity students were given capsules during Good Friday services. Ten students were given psilocybin while the other ten were administered a placebo (nicotinic acid). Psilocybin is the active ingredient in magic mushrooms, which have historically been used in spiritual ceremonies by a variety of native cultures. All of the divinity students were administered questionnaires that, in essence, asked about transcendental experiences (e.g., feelings of unity, transcendence with time and space, sacred feelings or a deeply felt positive mood). The authors concluded that statistically significant differences existed between the two groups. For example, four of the ten participants in the psilocybin group — and no controls — answered yes to at least 60 percent of the questions related to transcendental experiences.

In addition to psilocybin, a variety of other plant-derived substances such as mescaline, salvinorin A and tetrahydrocannabinol are known to distort intrapsychic perceptions and facilitate transcendental feelings. Mystical experiences can also be facilitated by emotional stress (Spilka et al. 2003, 317), temporal lobe epilepsy, rhythmic music and trance states (Spilka et al. 2003, 328). According to Newberg and others (2001, 43), electrical stimulation of limbic structures can produce "dreamlike hallucinations, out-of-body sensations, déjà vu, and illusions, all of which have been reported during spiritual states."

The variety of neurochemical stimulants behind transcendental experiences implies a certain biological substrate underlying religiosity. Scott Atran, author of *In Gods We Trust: The Evolutionary Landscape of Religion*, poses a few rhetorical questions to support the existence of a dedicated cognitive apparatus behind spirituality (2002, 14): "If religious belief and fantasy are pretty much the same, then why, for example, do individuals who are placed inside a sensory deprivation tank have an easier time evoking religious images than cartoon images, but not when those individuals are outside the tank?[1] Why do people in all societies spontaneously chant and dance or pray and sway to religious presentations, but do not rhythmically follow other factual or fictive representations so routinely?"

Studies using fMRI technologies are beginning to reveal the brain areas involved in religious thoughts (Kapogiannis et al. 2009a; Kapogiannis et al. 2009b; Azari et al. 2005). One of these studies had fifteen Carmelite nuns recall numinal experiences while the most active parts of their brains were mapped out. The nuns were asked to "remember and relive (eyes closed) the most intense mystical experience ever felt in their lives as a member of the Carmelite Order." As a control condition, the nuns were also asked to "remember and relive (eyes closed) the most intense state of union with another human ever felt in their lives while being affiliated with the Carmelite Order." In this experiment, a lot of cortical regions appeared to be activated by the recollection of mystical experience — right medial orbitofrontal cortex, right middle temporal cortex, right inferior and superior parietal lobules, right caudate, left medial prefrontal cortex, left anterior cingulate cortex, left inferior parietal lobule, left insula, left caudate, and left brainstem (Beauregard and Paquette 2006). One small problem with this study is that it only mapped out the recollection of

---

[1] The religious versus cartoon image observation is based on a journal article: Hood R. and R. Morris. 1981. Sensory isolation and the differential report of visual imagery in intrinsic and extrinsic subjects. (*Journal for the Scientific Study of Religion* 20:261–273.)

a mystical feeling and not the experience itself. Another fMRI study showed activations of the dorsolateral prefrontal, dorsomedial frontal and medial parietal cortex during recitation of religious material by devout members of an evangelical congregation (Azari et al. 2001). In both studies, the cortical areas activated are usually associated with abstract semantic processing and social cognition (Weissman et al. 2008; Iacoboni et al. 2004).

Something as simple as the number of certain neuroreceptors may fundamentally alter our sense of spirituality. In the only study of its kind, lower global densities of serotonin 1A receptors correlated with higher scores of self-transcendence (Borg et al. 2003). In other words, slight differences in the densities of a few neuroreceptors could influence whether we accept religion or not. The possible link between the serotonin system and religiosity is worth exploring, because serotonin is also implicated in the revelatory effects of psilocybin and LSD, as well as the pathophysiology of schizophrenia and mood disorders.

With so much evidence pointing towards a physiological and anatomical basis to spirituality, it should not come as a surprise that certain aspects of religiosity may be genetic. A number of studies have looked at possible hereditary factors behind religious attitudes, including one reared-apart twin study that found "genetic factors account for approximately 50 percent of the observed variance on our measures" (Waller et al. 1990).

Sociological Aspects of Religion

Not every distinguishable emotion is necessarily evolutionarily advantageous. Take, for example, déjà vu, the sense that one is re-experiencing an event that, in actual fact, has never happened before. The conjured memory of déjà vu can be so profound, disorienting and illusory that it seems to tamper with a person's fundamental sense of self. However, every facet of déjà vu suggests that the phenomenon is an accident of nature: déjà vu is rare, seemingly purposeless and does not alter any behaviors. Religious beliefs, on the other hand, are widespread and produce personal experiences that lead to group-level performances — elaborate rituals that possess plausible, if not compelling, evolutionary functions, especially in the context of hunting and gathering societies.

Recall that supernatural spirits are mysterious agents from beyond. Agents are entities (usually animals or persons) that appear to move in pursuance of their own goals. (Our predilection to divide the world into agents and non-agents is probably caused by a hardwired cognitive schema that predated religion.) Even though supernatural agents are inherently mysterious, traditional peoples tend to interact with spirits as if they are other human beings (i.e., agents). Spirits may be asked to provide protection from predators or reward hunting expeditions. They can be appeased, cajoled or even bribed in order to provide services like rain or good crops. Ancestor spirits, common in many tribes, can become morally outraged about social infractions like adultery or stealing. In hunting and gathering societies, spirits are not usually perceived as being absolutely omnipotent or omnipresent. For example, certain spirits may only be around at night or in specific locations (e.g., caves); therefore, they can potentially be avoided or even sometimes fooled. Yet, spirits characteristically possess more strategic information about any given situation than mere mortals do. According to Pascal Boyer, a spirit's superior hold on strategic information is an important distinction between supernatural entities and human beings (2001). Thus, a spirit is a fallacy but not a random fantasy — there is a very specific cognitive form and potential sociological function to these beliefs.

Coupling religion to ritualistic behaviors would evolutionarily maximize its effectiveness (Sosis 2004; Turbott 1997). Accounts from pioneering anthropologists suggest that life in hunting and gathering societies teemed with rituals, and most of them had spiritual overtones. There were ceremonies involving birth, initiation rites, marriage, death, healing the sick, locating important items, building canoes, constructing houses, controlling the weather, and war preparations, to name just a few. According to Sosis and Alcorta (2003), "Anthropologists and ethnologists have independently reached several common conclusions about ritualized behavior, most notably that it is a form of communication. The recurrent components of ritual, including exaggerated formality, sequencing, invariability and repetition, have been selected to facilitate communication by eliciting arousal, directing attention, enhancing memory and improving associations." In fact, Sosis and others did a methodical study of sixty randomly chosen societies and showed that the relative extravagance of male rites of passage correlated with the frequency of warfare (2007). Rituals allow the varying, nebulous emotions scattered through a community to be transformed into one unified position.

Given what we know about religion in its natural state (i.e., in hunting and gathering societies), an evolutionary function cannot be dismissed. After all, when a tribe spends a lot of time, effort and resources on an activity, it is likely to be evolutionarily advantageous (especially if the behavior appears to be instinctual, as opposed to logical). Although spirituality may be personally comforting, it is doubtful that the primary evolutionary purpose of religion is to lessen existential angst — surely evolution could find ways to allay anxiety that wouldn't involve the collective efforts of an entire tribe. Besides, atheists tend to do just fine. In fact, most experts who study the evolution of religion do not ascribe to the idea that it exists to provide personal comfort. My own survey of the literature reveals three possible evolutionary purposes to religion: it enforces altruistic behavior, it allows tribes to enjoy the benefits of divination and it promotes success in warfare.

Enforcement of Altruistic Behavior

The notion that morality and religion are fundamentally linked has been around for centuries. In the early 1900s, the pioneering French sociologist Émile Durkheim (1912) wrote about the societal aspects of religion in his definition: "A religion is a unified system of beliefs and practices relative to sacred things, i.e., things set apart and forbidden — beliefs and practices which unite in one single moral community called a Church, all those who adhere to them." Morality is a notoriously vague term that is perhaps better represented by the concept of altruism. Therefore, from an evolutionary perspective, morality can be framed as an ad hoc collection of socially altruistic attitudes and behaviors.

Both monotheistic and polytheistic religions include mechanisms that reinforce social contracts and formalize expectations of altruism (i.e., morality). Many hunting and gathering societies believe in ghosts or spirits that impose divine retribution for social infractions. David Sloan Wilson (2002, 23) provides an example from the Chewong peoples of the Malay Peninsula, who have many myths and superstitions around food sharing. It is commonly believed that those who flout the custom of conscientious food sharing put themselves under the constant specter of supernatural retaliation.

In *The Evolution of God*, Robert Wright describes a multitude of Polynesian gods who were always ready to pounce on those who commit various social infractions. Certain Tongan gods punish thieves by making them vulnerable to shark attacks, while other spirits bring bad luck to fishermen who argue with their spouses. Wright concludes, "When you add up all the little ways Polynesian religion encouraged self restraint, you wind up with a fair amount of encouragement — enough, perhaps to compensate for the absence of a centralized legal system" (p. 57).

The Benefits of Divination

The societal benefits of divination — seeking knowledge about the future — may be another characteristic of religion which has evolutionary value (Tanner 1978). Divining practices, in their various forms, appear to be universally present in hunting and gathering societies. Water witching is a divination act that attempts to locate underground sources of water by detecting the movements of a forked stick as it is held above the ground.

Another form of divination involves Cherokee shamans who monitor the drift of "two needles afloat in a small creek-side pothole or bowl of water" in order to come up with a proper cure (Lyon 2004, 129). Still another example involves the Maasai and Samburu shamans of northeast Africa who throw stones from hollow gourds or cow's horns. The resulting configurations portend various events, including good or bad fortune (Fratkin 2004, 207).

Of all the divination practices, scapulimancy has received the greatest attention from anthropologists. It involves the burning of a flat bone, usually the scapula of a deer or tortoise, and the analysis of the resulting cracks. The arrangement of the fissures are interpreted by shamans in order to choose such things as a new direction to hunt, a marriage partner (Park 1963) or type of medicinal cure (Wright 2009a, 73). Scapulimancy is practiced on several continents (Asia, Europe, Africa and North America), but it is not known whether it spread culturally or was independently invented at each separate location. (This latter possibility would truly be interesting.)

Omar Moore (1957) studied scapulimancy among the Naskapi peoples of Labrador and noticed that divination was often used to establish a new hunting direction after a series of failed expeditions (Park 1963, 198). He concluded that reading the cracks in a burnt shoulder blade gave essentially a random result, which could help tribes discover new solutions instead of continuing the same unsuccessful behavior — persisting to hunt in places that were no longer fruitful. Scapulimancy allowed everyone to agree on a new direction without long discussions filled with arbitrary arguments. A prophecy that was completely irrational — for example, a forecast to hunt in the direction occupied by a lake — would typically be overruled by consensus (and explained by, for example, comments suggesting the spirits were purposely misdirecting the shaman).

The essential feature of divination is that it produces quasi-random solutions to intractable problems. When one has absolutely no information to go on, a random solution is perhaps better than a misguided one. Each participant is unaware that the ritual leads to a randomly derived answer but, perhaps through evolution, those tribes that "blindly" used divination under certain circumstances had a survival edge. The power of divination allows an entire tribe to back an arbitrary decision — it is better to roll the dice and do something than not do anything at all. In the case of water witching, divination provides the tribe with a rallying point so that everyone digs for water, together, at the same location (Steadman and Palmer 2008, 140). Divination practices were almost always conducted by shamans, and consequently, each ritual became imbued with supernatural trappings. In fact, it may have been the disordered thinking of the shaman that helped inject randomness into such affairs.

Success in War

A number of religious scholars have noted the prominence of predator-prey themes in religious myths. The predator-detection cognitive module, if there is such a thing, might preferentially interact with religiosity. Although monotheistic religions see God as a sympathetic omnipotent figure, prehistoric religions tend to be more fixated on the nasty side of spirits. Scott Atran suggests "the evolutionary imperative to detect rapacious agents favors emergence of malevolent deities in every culture. Worship of serpent deities and would-be destroyers is at least as prevalent as God the father and mother goddesses" (Atran 2002, 79). Atran further argues, "Among the pre-Columbian Maya and Mexicans, for example, there appears to have been no entirely benevolent deity, and all were feared (to greater or lesser degrees) for their ability to bring death on almost anyone, almost anywhere" (Atran 2002, 75). Prehistoric religions all over the world fixate on predatory species — the tiger in Asian myths, eagles and coyotes in North American Aboriginal stories, jaguars in the Amazon jungles, and the rainbow serpent in Australian Aboriginal mythology, to name a few. Thomas Ellis has traced this theme from prehistory into modern religions (personal communication July 15, 2008; Ellis 2009) and proposes that the evil eye — a religious myth persistent in a number of ethnic groups — is a throwback to the predator's eye. Pascal Boyer circumspectly sums up this topic: "The connection to a predator-avoidance system may explain some of the emotional overtones of the religious imagination..." (Boyer 2001, 148).

Throughout recorded history, religion has repeatedly — perhaps paradoxically — been connected to warfare. Starting with the first ancient city-states and proceeding to modern nations, religious zealotry and warfare frequently coexist. The pursuance of war to fulfill Aztec human sacrifice rituals is perhaps one of the most gruesome examples. The wars of ancient Mesopotamia, the Crusades, the Muslim conquests and many of the European wars during the sixteenth and seventeenth centuries were inextricably linked to religious beliefs. According to Robert Wright, even the roots of Christianity may have more to do with hostile tribalism than with staking out a moral place in the universe (Wright 2009, 103): "And if you go back to the poems that most scholars consider the oldest pieces of the Bible, there's no mention of God creating *anything*. He seems more interested in destroying; he is in large part a warrior god."

Allen D. MacNeill (2004) straightforwardly connects the evolutionary purpose of religion to warfare in an explicitly titled article, "The Capacity for Religious Experience is an Evolutionary Adaptation to Warfare." MacNeill makes a cogent argument that religion has adaptive qualities in traditional societies and then makes the following proposal: "By making possible the belief that a supernatural entity knows the outcome of all actions and can influence such outcomes, that one's 'self' (i.e., 'soul') is not tied to one's physical body, and that if one is killed in battle, one's essential self (i.e., soul) will go to a better 'place' (e.g., heaven, valhalla, etc.) the capacity for religious experience can tip the balance toward participation in warfare."

In hunting and gathering societies, warfare was both common and exalted by shamanistic practices. In traditional societies, the decision to go to war usually rested on the chief and his inner circle, but with consultation with the shaman. Once the decision to go to war was made, religious rituals invariably accompanied war preparations. Furthermore, shamans were often asked to predict the outcome of impending conflicts — a situation that would have created powerful evolutionary selection pressures. In fact, several anecdotes exist (which I will review in Chapter 8) of shamans blaming neighboring tribes for tragic events, a scenario which would have increased the probability of preemptive attack.

Although it is undeniable that hominids have, to some degree, evolved for war, scholars do not agree on the most important factors driving people into group conflicts. In hunting and gathering societies, the reasons for war are likely multifactorial and I doubt if they are ever fully understood by visiting anthropologists. Studies on traditional war have implicated territorial disputes for access to resources, as well as revenge for homicides (Keeley 1996).

Regardless of the reasons for war, spirituality is integrally involved in all forms of intertribal conflict, suggesting that shamans had a disproportionate input into the group decision for warfare. Because warfare so often results in many deaths, the spiritual practices of each shaman could have been evolutionary game-changers. In other words, of all the shamanistic endeavors, pre-war attitudes (and rituals) could have been the most evolutionarily important. As a consequence, hominid warfare may have created the greatest selection pressures guiding the development of shamanism and religion.

Religion and Group Selection

A number of experts agree that religion forces people to commit to the needs of the tribe as opposed to their own selfish interests. Imbuing religion with complex rituals and initiation rites produces a "hard-to-fake" and therefore honest signal of commitment, meant to dissuade freeloaders who may be tempted to put their needs ahead of the group (Atran 2002; Sosis 2004; Steadman and Palmer 2008, 29). Although most hypotheses about the altruism of religion do not mention group selection, I am not convinced that such arguments can avoid this evolutionary mechanism.

David Sloan Wilson (2002) proposes that religion is a group-selected behavioral trait. He begins with this premise: "Religions are often concerned with the necessities of life — food, shelter, health, safety, marriage, child development, social relations of all sorts. These are so obviously related to survival and reproduction...." He also emphasizes that "religions are well known for their in-group morality and out-group hostility...." Taken together, there is an implication that religion is a group-selected behavioral trait. Wilson also deals with an interesting loose end about religion: how can fallacious beliefs be evolutionarily adaptive? He suggests that even fantasies can be adaptive, so long as those beliefs result in adaptive behaviors — something that religion does.

So how does atheism fit into this evolutionary picture? Atheists seem to be a minority in most contemporary societies. The following are the percentages of people who identify themselves as atheists in some Western nations: U.S: 3 percent, Canada: 9 percent, France: 19 percent, Sweden: 17 percent and England: 10 percent (Hunsberger and Altemeyer 2006). Studies on the psychological profiles of atheists provide no evidence that atheists are more immoral or selfish than their religious counterparts. According to systematic studies, atheists in contemporary societies tend to be less prejudiced, less dogmatic and less vulnerable to in-group versus out-group thinking (Hunsberger and Altemeyer 2006, 128).

Wilson suggests that the evolutionary presence of atheists reflects a group-selected dynamic whereby tribes are best served when there is a healthy debate between spiritual ideas and practical suggestions (Wilson 2002, 229). Alternatively, the term *atheist* may not be entirely apropos when discussing hunting and gathering societies. Atheism implies the jettisoning of every single folk story or superstition — something that is much easier to do in a monotheistic culture. The anthropological literature reveals that shamanistic prophecies are sometimes met with outright skepticism and disbelief, but I have never read an account of a tribesperson questioning every single quasi-spiritual idea in their belief system. A certain amount of skepticism inevitably runs through every society; however, atheism per se may not be an evolutionarily relevant phenotype in traditional societies.

Conclusion

Academia seems to be torn about whether religion is an evolutionary byproduct of other contiguous cognitive abilities, or an evolutionarily advantageous construct in its own right (Dawkins 2006). In my view, there is substantial evidence suggesting that spirituality is the child of evolution. First, when animals spend a substantial amount of time and resources engaged in a certain activity, that behavior is usually evolutionarily adaptive. Second, a number of evolutionarily positive behaviors spring from religion, especially in the context of hunting and gathering societies. Religion is complex and may serve more than one evolutionary function, including facilitating warfare, divining, or reinforcing altruistic behaviors. Some aspects of religion could be incidental byproducts, but it seems unlikely that religion as a whole is evolutionarily neutral.

The religions of traditional societies enhance social cohesion while simultaneously being concerned with far-away dangers. In hunting and gathering societies, spiritual beliefs frequently include themes of intangible otherworldly dangers. This is not a coincidence: monitoring danger is a very rudimentary animal behavior. The dangers feared by tribal religions are fictitious but the constant concern perhaps encourages a level of vigilance that prejudices neighboring tribes and serves as a constant reminder of their potential deadliness.

Tribes may require mechanisms like spirituality to achieve the proper balance between aggression, amity and vigilance to contend with neighboring tribes. Surprise raids, because of their relative infrequency, would be difficult to defend against: years of freedom from attack could create complacency and lower vigilance in a tribe. (A contemporary analogy may be how most of us have, perhaps inappropriately, accommodated to the threat of nuclear devastation or global warming.) Religion may have been useful to counter the tendency of human groups to habituate to an infrequent but devastating threat: a surprise attack from another tribe. Because tribal attacks have probably happened for several million years (and result in a large percentage of deaths), the selection pressures they imposed were significant. Using spiritual themes, shamans — the creators of religion — could have conjured up the fear of threats from neighboring tribes, which ultimately made those risks tangible and kept them in the forefront of everyone's mind. Put this way, it's perhaps a wonder that nature did not evolve something like religion any sooner!

– 8 –
Shamanism

The term *shaman* is derived from a word meaning "he who knows" in the language of the Evenki (formerly known as the Tungus), an indigenous people of southern Siberia (Lindholm 1990). What these shamans "knew" was the landscape of the ethereal spirit world. Europeans described the intriguing practices of Siberian shamans as far back as the late 1600s, but it was during the late 1800s that shamans began to get the attention of trained ethnographers, and it was then that comprehensive descriptions of their rituals were laid down (Znamenski 2004, 2007).

Defining Shamanism

The strictest definition of shamanism refers to those spiritual practitioners from Siberian nomadic cultures, but this narrow definition is outdated. Most anthropologists agree that Siberian shamans are nearly identical to spiritual practitioners from other societies: medicine men, diviners, sorcerers, magicians, jugglers, witch doctors, exorcists and mediums. Some experts in shamanism use the term *magico-religious practitioner* to describe any spiritual leader affiliated with traditional societies (Winkelman 2004). This expression is perhaps the most descriptively accurate, but the term has not really caught on, and so I will refer to shamans and magico-religious practitioners interchangeably.

Shamans are believed to possess spiritual powers involving communication with a world beyond the observable realm. Because shamans act as conduits to a greater spirit world, they are seen as the religious leaders of their communities. One of the most notable roles for shamans is endeavoring to heal the sick (McClenon 2002; Vitebsky 1995). They often lead elaborate rituals meant to rid the body of malevolent spirits, the usual supposed reason for illness in tribal societies (Narby and Huxley 2001). A shaman normally presides over rites of passage such as birth, coming of age, marriage and death. Rituals involving a community's procurement of food, such as rain dances or divining the movement of prey, are also under a shaman's purview, as are ceremonies in anticipation of war.

Universality of Shamanism

In the early 1700s, thousands of miles from Siberia, a French Jesuit missionary living among the Huron and Iroquois of eastern North America chronicled how foreign spirits appeared to regularly possess the local magico-religious practitioners. Lacking any established terminology, he called them *jongleurs* or jugglers, implying they were magicians. He said of these jugglers, "...there is nothing outside the scope of their knowledge. Predictions of the future, the success of a war or of a journey, the secret causes of a malady, bringing good luck to a hunting or fishing party, finding stolen objects, casting spells and curses — in short, everything that concerns divination — is absolutely within their jurisdiction" (Narby and Huxley 2002, 24).

Although our French Jesuit missionary appears to have been unaware of the concept of shamanism, his description of the Iroquois jugglers is unmistakable. This highlights one of the most astounding features of shamanism: its universality. In fact, some variant of magico-religious practice has been found in every hunting and gathering society. In more complex societies, traditional shamanism evolved into societal roles such as medicine man, diviner, witch doctor, medium, healer and other types of magico-religious practitioner. An analysis of forty-seven societies throughout the world, dating back to 1750 BC, revealed that each one of them included some form of magico-religious practitioner (Winkelman 1990).

The notion of being possessed or tangibly connected to spirits seems patently absurd, yet such ideas have been observed in all corners of the globe. In my view, the utter preposterousness of magico-spiritual beliefs is the smoking gun that implicates a genetic foundation to shamanism. Magico-religious ideas are neither a logical response to any basic human need nor are they a rational solution to any social predicament.

The spiritual pronouncements associated with shamans carry distinctive and universal themes. They can arguably be characterized as a sort of multifaceted cognitive reflex (i.e., emotion or fixed action pattern) similar to love, jealousy or anger. Recall that in contrast, the control of fire, which is also universally observed in traditional societies, is so undeniably useful that no prompting by evolution is required to explain it.

Just as thirst prompts a search for liquids but culture influences the drink you grab, the inclination to delve into the spiritual realm is constitutional but the specific details vary. I propose that the general themes of shamanistic jabber, which so often originate in psychosis, are ultimately programmed by evolution. The expression of *de novo* spiritual musings (i.e., psychotic thoughts) helps propagate religion and ultimately enhances survival in traditional life.

Several anthropologists have recognized that specialized behaviors associated with shamanism could be governed by genetic determinants and therefore might be evolutionarily driven. According to shaman expert Michael Winkelman (2004), "The worldwide distribution of the shaman in widely separated societies is not a consequence of diffusion, indicating that the source of shamanism is an independent invention and human psychobiology." Noll (1989) also considers an innate component to shamanism based on Shirokogoroff's observations that shamanism is so frequently associated with some degree of mental instability. Still others have proposed that shamans may be susceptible to hypnotic trances or even have constitutional "fantasy-prone personalities" (Wilson and Barber 1981, 1982, cited in Noll 1989)

Here is where the psychological sciences might assist the anthropologists — by answering the following crucial question: Of all the psychological qualities characteristic to shamans, which trait is under the greatest genetic influence? There are dozens of possible psychological traits associated with shamanism: hypnotic susceptibility, creativity and an inclination towards spirituality. But the best candidate may be schizophrenia because it most thoroughly resembles shamanism while also being highly heritable. If one strips away the cultural accouterments associated with shamanism and compares it to the fundamental expression of schizophrenia, the two look very much alike.

One interesting fact about shamanism is that, just like mental illness, it is often perceived as something that overcomes an individual. In South Africa, for example, the Thonga people divide medicine men into two groups, "ordinary" and a special group who have suffered possession by a spirit (*bubabyi bya psikwembu*, which according to Ackerknecht literally means "the madness of the gods") (Ackerknecht 1943). One of the first documented accounts of an involuntary shamanistic calling comes from an acculturated Tlingit shaman, George Hunt, whose story has survived because Hunt became an assistant to the renowned nineteenth century anthropologist Franz Boas (Znamenski 2007).[1] For almost a year, Hunt had felt unwell; he had to be restrained because he was acting "like someone wild." Hunt surmised he had visited another world and as a consequence, his relatives concluded that certain supernatural powers had entered his body — a clear signal that he should become a shaman. This attitude — that a person afflicted with insanity must answer the call to become a shaman — is evident throughout the world. Other cases can be found in Siberia (Eliade 1964, 13; Czaplicka 1914, 9; Vitebsky 1995, 60), India (Narby and Huxley 2001, 115), Australia (Eliade 1964, 45), Africa (Junod

---

[1] Andrei Znamenski (2007, 61) reasonably speculates that Hunt's identity as a shaman was originally obscured by Boas because of the Canadian government's prohibition on shamanism.

1962, 514; Ackerknecht 1943, 69), North America (Devereux 1961, 14; Irwin 1994, 88) and South America (Ritchie 1996).

Psychosis in Shamans

Michael Winkelman (1989, 1990) has analyzed the role of shamans across cultures and his succinct yet comprehensive portrayal highlights how psychotic-like experiences are a salient feature of shamanism (1989, 19):

> Shamans are selected and trained through a variety of procedures and auguries, including having had involuntary visions, having received signs from spirits, having experienced serious illness, having deliberately undertaken vision quests, and having induced trance states through a variety of procedures, such as hallucinogens, fasting and water deprivation, exposure to temperature extremes, extensive exercise (e.g., dancing and long distance running), various austerities, sleep deprivation, auditory stimuli (e.g., drumming and chanting), and social as well as sensory deprivation. Their trance states are generally labeled as involving soul flight, journeys to the underworld, and/or transformation into animals.

A perusal of the anthropological literature on shamanism reveals a repeated pattern of psychotic experiences — commonly auditory hallucinations — occurring unexpectedly in young people soon to become shamans (Narby and Huxley 2001; Vitebsky 1995; Grim 1983; Irwin 1994; Ritchie 1996; Devereux 1961; Eliade 1964). The posited connection between shamanism and psychosis has been obvious to some anthropologists and doubted by others. Over the next few pages, I will provide several examples of psychosis in magico-religious practitioners, and evaluate how the perception of shamanistic practices has vacillated over the twentieth century.

The idea that shamanism is associated with mental instability (or even complete insanity) was a recurring theme of the nineteenth century anthropologists who immersed themselves in the traditional cultures of northern Russia, such as the Chukchee, Yakuts and Evenki. A male shaman called Scratching-Woman became the archetype of the "half-crazy" shaman in the early twentieth century (Bogoraz quoted in Znamenski 2007, 80). Scratching-Woman was described by anthropologist Vladimir Bogoraz (also called Waldemar Bogoras) as being excitable, irritable, deceitful, quarrelsome and on the "verge of insanity." Bogoraz was not a psychiatrist and did not present any convincing evidence that Scratching-Woman was truly psychotic. (Nowadays, Scratching-Woman would likely be diagnosed with bipolar II disorder.) However, Bogoraz did observe several other Chukchee shamans showing the telltale signs of psychotic illness. For example, he documented, "I have before spoken of the female shaman Te'lpiñä, who, according to her own words, had been violently insane for three years, during which time her household had taken such precautions, that she could do no harm to the people or to herself" (Bogoraz-Tan 1904). Another shaman told Bogoraz that, in his youth, a voice had told him to go into the wilderness and find a "tiny drum," which he supposedly did and ascended into the sky and pitched his tent on the clouds. Bogoraz's conclusions were that most shamans were peculiar and unpredictable (Bogoraz-Tan 1904).

Another anthropologist, Wacław Sieroszewski, recounted the testimony of a Yakut shaman in the late nineteenth century: "When I was twenty years old I became very ill and began to see with my eyes and hear with my ears that which others did not see or hear. For nine years I attempted to overcome my illness and did not tell anyone what was happening to me, since I was afraid that people would not believe me and would laugh at me. Finally I became so ill that my death seemed unavoidable. When I began to perform shamanistic rites I became better; and now if I do not exercise my powers for a long time I am not well, I become ill!" (Sieroszewski 1993).

In 1935, Sergei Shirokogoroff published a book entitled *Psychomental Complex of the Tungus* which contained several detailed accounts of shamanism. Because he was a physician, Shirokogoroff attempted to provide diagnostic insights about these curious shamans from northern Russia. Shirokogoroff described four cases of shamanistic initiation, two of which contained possible psychotic episodes. (A third account (Case 2) may have been based on a psychotic shaman, but the story is so incredible that it reads more like a folktale.)

In the first case, the community believed a spirit had entered a certain woman. Shirokogoroff observed this case and wrote, "she trembled, ran away, climbed up a tree, refused food, 'forgot' everything" (p. 346). During this period the woman disappeared in the woods for eight days; however, upon her return, she claimed she never left. Preparations were subsequently made for her to become a shaman.

Another case (Case 3) described a woman having "suspicious fits in which she was visited (entered) by the soul of her lover..." (p. 347). She, too, trembled, climbed trees, ran away into the forest and refused food. During this period of aberrant behavior, the afflicted woman couldn't or perhaps wouldn't recognize any one around her. Shirokogoroff concluded, "Her cognition of reality was rather doubtful."

Shirokogoroff was skeptical that such behavior reflected veritable psychiatric pathology, and instead classified it as hysterical reactions.[2] He noticed that the frenzied people frequently disappeared into the forest during the summer, as opposed to in the dead of winter when such actions would be life-threatening. According to Shirokogoroff, this hint of voluntary behavior ruled out authentic insanity. Shirokogoroff, however, was making a mistake frequently made by observers unfamiliar with psychotic individuals. The fact that a psychotic person can modify his or her behavior does not exclude the possibility of a psychotic process. Psychotic patients on hospital wards are routinely directed by health care staff and encouraged to abide by social norms. The complete inability to modify behavior only occurs in the most severe psychotic episodes.

Shirokogoroff also described five practicing shamans of which four showed no evidence of unusual behaviors. One shaman may have been a little unusual and might have possessed mild psychotic symptoms. However, even if we take Shirokogoroff's observations at face value, it should be noted that when it came to selecting shamans, the Evenki were known to be especially choosy. For the Evenki, possession by spirits (i.e., psychosis) was just one of many assets used to select shamans. Intelligence, honesty, and knowledge of shamanistic rituals were also taken into account.

In 1943 E. H. Ackerknecht, who was both a physician and medical historian, published one of the first cross-cultural analyses of psychopathology. His paper focused on the many aberrant behaviors so often associated with shamans (including those described in Shirokogoroff's work). Ackerknecht saw a global pattern emerging and supplied this account of a prospective Chukchee shaman as a prototypical example (Mikhailovsky quoted in Ackerknecht 1943):

> He who is to become a shaman begins to rage like a raving madman. He suddenly utters incoherent words, falls unconscious, runs through the forests, lives on the bark of trees, throws himself into fire and water, lays hold on weapons and wounds himself, in such wise that his family is obliged to keep watch on him. By these signs it is recognized that he will become a shaman....

Based on this and other similar accounts, Ackerknecht suspected that most shamanistic initiations, previously felt to be hysterical reactions, actually described schizophrenia.

It would only be a matter of time until psychiatrists found themselves weighing in on the question of whether shamans were psychotic. One of

---

[2] "Hysterical," in this case, meaning characterized by an excessive or uncontrolled outburst of emotion. Hysteria isn't a diagnosis in psychiatry, but when used implies the behavior is not caused by psychosis.

the first psychiatrists with anthropological skills was George Devereux, who studied the Mohave people around the 1930s and supplied a number of detailed psychiatric histories in his book *Mohave Ethnopsychiatry and Suicide: The psychiatric knowledge and the psychic disturbances of an Indian tribe* (1961). He was convinced that most shamans he had met were "mentally deranged" and many were outright psychotic (Narby and Huxley 2001, 120). For example, Devereux described the case of a forty-six-year-old Mohave shaman admitted to a state psychiatric facility in Arizona in the mid-1930s. The patient was described as peculiar, delusional, staring into space and uttering nonsensical comments. According to a reliable member of his native community, "When he is excited he mutters abruptly in English, 'Go West, and Go West, and to the East, with a pick and a shovel, over 2000, over 2,000,000 (repeats the numbers).'" The informant added, "Not until he went crazy did people know he was a shaman, but now that he practices, he is better. He cures colds, pneumonia, and the flu. He began practicing about a year ago, when he was released from the asylum." The hospital discharge diagnosis was "Manic-depressive psychosis, manic type."

Another psychiatrist in regular contact with traditional societies during the early twentieth century was B. J. F. Laubscher (1937), who documented over one hundred cases of schizophrenia among the Thembu and Fingo tribes of South Africa. His 1937 book *Sex, Custom and Psychopathology: A study of South African Pagan Natives* is replete with examples of psychosis in people affiliated with shamanism. Laubscher was primarily a psychiatrist and did not appear to have any preconceived opinions about whether shamans were psychotic. (In fact, he never used the term *shaman* and instead referred to these magico-religious healers as witch-doctors.) According to Laubscher, only the most severe instances of psychosis among the Thembu tribes would ever present to European hospitals. These cases were either a nuisance or perceived danger to the nearby Europeans or unmanageable psychotic people brought in by the natives themselves (e.g., those accused of assaults on children or burning down huts). Laubscher observed, "It will be realized that the demands for adjustment in his natural environment are so simple that that the average patient can continue to live in the kraals without seriously disturbing the routine of life. Besides, if his psychotic abnormality assists him to function as a herbalist or witch-doctor, he becomes quite an economic asset to his family"[3] (p. 226). Just like Devereux who studied the Mohave peoples, Laubscher was convinced that many of the South African tribal witch-doctors were "psychotic persons in remissive phases" (Laubscher 1937, 227).

---

[3] The word *kraal* is used in this case to mean a village or settlement.

Laubscher's clinical perspectives dovetailed with the descriptions of shamans provided by anthropologists studying other South African tribes. The Swiss anthropologist Henri A. Junod (1962) wrote extensively about his (early twentieth-century) contact with several South African Thonga tribes. He described one case of a Thonga man whose induction to exorcism had been preceded by an apparent psychotic episode. The exorcist recounted how his ancestor spirits had earlier possessed him. Others had told him that he behaved erratically. Over a six-month period, he became "lame" and lost his appetite. He was also partially amnestic during this period — he did not recall having attacked people without any apparent reason. The other community members concluded, "he is sick from the gods" (p. 481). Another Thonga man characterized as an "exorcist, witch-doctor, medicine-man and magician of Heaven" may have also been experiencing hallucinations and paranoid delusions when he said, "At noon when the sky is quite clear, I see a little cloud appearing over my head, quite black. The lightning bird is in it, and wants to kill me, having been sent by one of my enemies" (p. 519).

In all corners of the globe, visiting anthropologists chronicled similar experiences reported by shamans. In Polynesia, the anthropologist E. M. Loeb (1924) confirmed that shamans were responsible for healing the sick, prophesying and placing curses on the enemy. Most Polynesian shamans appeared to have had histories of epilepsy or "temporary insanity." Loeb described an elderly Polynesian shaman he visited in 1923 who would sometimes see ghosts sitting atop graves. He explained that angry ghosts could enter anyone, but they did not address ordinary people (p. 400). Loeb added, "My informant is in the custom of communicating with his dead elder brothers, Fatamaka and Haimatatau. He actually sees them" (p. 399).

After World War II, traditional life began to fade throughout the globe, yet descriptions of shamanism persisted. Based on fieldwork in Nepal during the early 1970s, a typical shaman of Kham Magar ethnicity was described in the following way (Watters 1975, 126):

> Bal Bahadur Jhankri, a shaman of Taka village, was first visited by the spirit of his dead grandfather when he was 14 years of age, five years after his grandfather's decease. During this time he became very ill, losing his appetite and relish for food. He describes himself as being "like a mad man" for a full year, not knowing where he was, or what he was doing. He withdrew to the mountains and experienced long periods of unconsciousness, during which time his soul wandered in the four directions of the compass. He learned from the gods, and from his ancestor spirit, the nature of diseases and their cure,

and also how to fight demons. He had many encounters with evil spirits and demons who attacked him with knives, cutting out his heart and liver. He fought back, poking at them with burning fire brands, but to no avail. His tormentors always won in the end.

In the Amazon jungle of southern Venezuela, a Yanomamo shaman named Jungleman disclosed, "After I became a young man, everything in the jungle talked to me. A log once asked, 'Why are you stepping on me?' " (Ritchie 1996, 21) and at other times, bird whistles sounded like people talking (p. 23). The shaman further recounted how he had visited the legendary folk spirit Omawa and had felt the presence of an unfriendly enemy spirit, Yai Wana Naba Laywa (p. 25).

In the early 1960s, a Japanese psychiatrist conducted one of the most ambitious studies of shaman-like people to date. Yuji Sasaki (1969) interviewed dozens of mediums, fortune-tellers and priest-equivalents from a number of Japanese rural communities. Sasaki provided detailed histories using contemporary psychiatric terms such as *delusions of possession*, *psychotic grandiosity*, *auditory hallucinations*, *loose associations*, and *incoherence*. 29 percent (sixteen of fifty-six) of these magico-religious practitioners reported spontaneous psychotic symptoms preceding their conversion to becoming professional spiritualists. Of the fifty-six subjects, six had diagnoses of schizophrenia (11 percent).

I have admittedly cherry-picked examples of possible shamanistic psychosis. There are many more vignettes that might reflect psychosis in shamans, but too little supplementary information is provided in the majority of anthropological accounts. This lack of substantive narrative hinders any reasonable attempt to differentiate between psychosis and ordinary shaman-speak. For example, a Winnebago medicine-man claimed he had been dead, as well as previously transformed into a fish (Radin 1937, 115); an Inuit shaman professed to be telepathically informed that his relative had died (Frederiksen in Valentine and Vallee 1968, 52). Such accounts may well be genuine psychosis, but they could also be the equivalent of your neighborhood preacher repeating the folktale about Jesus walking on water.

The Epidemiology of Shamanism

Shamanism and schizophrenia share many more features than just psychosis. For example, the onset of schizophrenia during young adulthood and the intensification of symptoms during periods of stress parallels shamanism. The fact that males are more severely affected than females with schizophrenia is also consistent with shamanism. (There are female shamans but they tend to be less common in most traditional societies (Tedlock 2005).) Anthropologists have noticed that shamanism tends to run in families, although not always, which is in accordance with the genetic transmission profile of psychotic disorders. Similar to patients with schizophrenia, shamans tend to prefer social isolation.

Another interesting comparison is that some shamans have been noted to have "maniacal" eyes, similar to the facial expression sometimes seen in grossly psychotic persons (Czaplicka 1914, 11). Last, schizophrenia's 1 percent prevalence corresponds well to the average population count of hunting and gathering tribes. Because hunting and gathering tribes normally contain between 60 and 150 individuals, a typical tribe would usually be ensured one psychotic-prone shaman.

Shamans and Altered States of Consciousness

A shaman endeavors to interface with and manipulate the supernatural world in the service of healing the sick, prophesying, divining, and negotiating rites of passage. Making a connection with the illusory spirit world can be accomplished through a variety of methods, such as telepathic communication, symbolic transformation into animals, vision quests, soul flight, or spirit possession. During shamanistic ceremonies, shamans often appear to fall into an altered state of consciousness (Harner 1980; Krippner 2002). This state has been described by many terms: ecstasy, dissociative state, auto-hypnosis, hysteria, trance, possession trance, spirit possession, magical flight, soul journey, soul flight, and out-of-body experience. Given the variety of terminology applied, it should come as no surprise that there has been much controversy on how to precisely characterize the shamanistic state of mind (Bourguignon 1989).

Trance may be the most accurate term to describe a shaman's mental state during ritualistic performances. A trance is a type of altered state of consciousness — a feeling of reality different from usual experience. This description may seem unsatisfactory, but because defining consciousness is so problematic (Jaynes 1976), outlining the specific parameters of any altered state turns out to be equally elusive.

Trance states come in various forms — for example, some are placid while others frenzied. Memories of trance states also vary, from vivid recollections to complete amnesia. (I have hypnotized a few patients in medical settings and all of them have claimed a vague recollection of certain portions of the experience.) Western civilization offers few opportunities to experience trance-like phenomena. In addition to formal hypnosis, trance-like states can sometimes be reached during fringe religious ceremonies, meditation exercises or through the use of drugs like marijuana. Even musicians jamming late into the night, filling their senses with loud pulsating sounds, will sometimes feel a light trance. It is generally assumed that given the right circumstances almost anyone has the ability to enter a trance. (Incidentally, the hypnotizability of patients with schizophrenia has been found to be comparable to the general population (Frischholz et al. 1992; Murray-Jobsis 1991; Pettinati et al. 1990; Jamieson and Gruzelier 2006; Kramer and Brennan 1964).)

One critical debate centers on whether a shaman's altered state of consciousness is a voluntary hypnotic-like trance or an involuntary "spirit possession" — the latter more closely resembling a true psychotic episode (Noll 1989; Wright 1989; Peters and Price-Williams 1980). According to shamanism expert I. M. Lewis (2003, 48), choosing between these two alternatives may be unnecessary, as both psychological states seem to be recognized by traditional societies. The Evenki, for example, were known to distinguish between involuntary possession and voluntarily succumbing to spirit possession. Although it would be more convenient for me to side with the involuntary-spirit-possession camp, my impression is that most shamanistic ceremonies are voluntary trances. This, of course, does not preclude the presence of chronic low-level psychosis during shamanistic proceedings. Besides, the "initiatory possession" that so often inducts prospective candidates into shamanism appears to be genuine psychosis.

Shamans enter trances by a variety of methods including chanting, drumming, sleep deprivation or sensory deprivation. Some trance-inducing methods such as sleep deprivation and sensory deprivation are well known to also exacerbate psychosis in vulnerable people. In addition, shamans will frequently ingest hallucinogenic substances during rituals in order to alter their sense of reality. (Shamanistic hallucinogens will be explored more closely in the next chapter.) Therefore, some shamans may modulate both trance induction (e.g., chanting, drumming) and psychotic activation (e.g., sensory deprivation, hallucinogens). The ability to separately modulate these two intersecting conditions — hypnotic trance and psychosis — could create a spectrum of mental states, which would explain the varied presentation of shamanistic states of consciousness.

The idea of modulating psychosis may seem strange because contemporary medical thinking has been overly categorical regarding psychotic symptoms. Psychiatrists have created artificial boundaries to define when a certain thought or hallucination is permitted to enter the medical category of psychosis. Psychosis also tends to be framed as a medical affliction — one that utterly overwhelms a patient, like cancer or a heart attack. This analogy may be appropriate for severe psychosis, but milder forms of psychotic thinking better resemble conditions like tension headaches or irritable bowel syndrome, where the intensity of symptoms can be suppressed or exacerbated by a patient's immediate thoughts or behaviors. For example, psychotic patients will sometimes talk to themselves in the waiting room but then fully suppress that behavior during their appointment with the psychiatrist. (It is perhaps for this reason that CBT has recently been shown to be successful in mitigating hallucinations in schizophrenia (Wiersma et al. 2001; Morrison 2002).) Attempting to control and modify psychotic experiences has historically been associated with shamanism. For example, the previously mentioned Yanomamo shaman Jungleman, in discussing his own auditory hallucinations, declared, "I kept remembering my mother's words: 'As you learn to control them, they will stop scaring you.' How right she was..." (Ritchie 1996, 25).

Examples of Psychosis in Non-Shamans

The anthropological literature reveals many examples of apparent psychosis in individuals never destined to become shamans. But, even when psychosis strikes ordinary people in traditional societies, their ramblings tend to contain magico-religious themes. Examples of non-religious psychoses do exist (Brant 1969; Valentine and Vallee 1968; Znamenski 2007, 82) — one Inuit woman held the relatively simple belief that a fox was inside her — but these kinds of non-spiritual delusions are infrequent. In most cases, delusions and hallucinations contain references to ghosts and spirits. For example, during his stay with the Tapirapé people of Brazil in 1939–40, anthropologist Charles Wagley (1977) heard about several ghost sightings: "One woman saw a ghost bathing in the small brook when she went to fetch water after sundown. A man who remained alone in his garden to sleep saw the ghost of a person who had died over 10 years before."

Paul R. Linde (2002), an American psychiatrist working in Zimbabwe during the 1990s, made the following observation, "...Zimbabweans do not report hearing auditory hallucinations of Jesus Christ, rather they report hearing those of their ancestor spirits. They are not paranoid about the FBI, rather they are paranoid about witches and sorcerers" (p. 56). Linde's clinical observations are in keeping with studies that have sought to determine the nature of delusions and hallucinations in traditional communities (Teuton et al. 2007; Patel et al. 1995; Edgerton 1966). For example, after interviewing dozens of community workers, nurses and traditional healers, a study from the University of Zimbabwe medical school concluded, "Angered ancestral spirits, evil spirits and witchcraft were seen as potent causes of mental illness" (Patel et al. 1995). In New Guinea, the Bena Bena people believe that almost every illness, including insanity, is connected to sorcery or malevolent ghosts (Langness 1965).

People who appear to be targeted by spirits sometimes induce fear in others, but they can also engender reverence. As far back as the early twentieth century, scholars who studied prehistoric religions noticed "...mentally ill people are often regarded as holy in primitive societies" (Rank 1967). Similarly, during a 1920s expedition to the Polynesian island of Niue, anthropologist E. M. Loeb observed, "Feeblemindedness is treated with scorn in Niue today, but insanity still calls forth respect" (Loeb 1924, 402).

New fodder for magico-religious ceremonies does not necessarily start with shamans. When people in the community hear their deceased ancestors speaking or feel they are being watched by demons, the experiences keep religion front and center. These profound experiences will eventually be incorporated into the spiritual discourse of the community, which ultimately serves to further the shaman's agenda.

Dissenting Opinion

There are undoubtedly many shamans who lack a history of psychosis (Elkin 1977; Radin 1937, 111; Walsh 1997, 2007), yet to dismiss any association between insanity and shamanism is sweeping too much under the rug (Krippner 2002). Take Mircea Eliade's (1964) classic book on shamanism, originally published in French in 1951. From the outset, the manuscript was colored by Eliade's personal conclusions about the history of religion. Before writing the book, Eliade jotted a note to himself: "I must present shamanism in the general perspective of the history of religions rather than as an aberrant phenomenon belonging more to psychiatry" (Znamenski 2007, 170). Eliade was clearly working under the assumption that religion and insanity were unrelated, and so he felt his hand forced to choose the better option. Eliade suggested shamans were, for the most part, healthy individuals (p. 30–31) and yet anecdotal hints of psychosis repeatedly slipped into his writings. For example, Eliade wrote, "Among the Tungus of the Transbaikal region he who wishes to become a shaman announces that the spirit of a dead shaman has appeared to him in a dream and ordered him to succeed him. For this declaration to be regarded as plausible, it must usually be accompanied by a considerable degree of mental derangement" (p. 16). Eliade also distinguished between ordinary shamans and those contacted by the spirit world (which I interpret as probable psychosis): "But these 'self-made' shamans are considered less powerful than those who inherited the profession or who obeyed the 'call' of the gods and spirits" (p. 13).

The idea that psychosis has little to do with shamanism was also proffered by Jane Murphy (1976), who asserted that the Yupik word *nuthkavi-*

hak, meaning "being crazy," had never been used to describe any of the eighteen shamans in an Alaskan Yupik community of 499 people. During her fieldwork in 1954–55, Murphy also identified six persons believed to be witches but never referred to as insane. One conclusion is that Murphy's observations were accurate — perhaps she had witnessed an Inuit shamanistic culture with less insanity than most. Alternatively, Murphy's conclusions could be dismissed as semantics; the Yupik terminology may only distinguish severe acute psychosis from its milder forms. My suspicion is that both of these explanations have merit. Among traditional societies, Inuit culture may be unique in the sense that spirit possession (i.e., insanity) never became integrally associated with ancestor-worship (i.e., religion); this sets it apart from most traditional societies (Radin 1937, 222). Instead, shamanistic aptitude among the Inuit was derived from perceived magical power (Carpenter 1968). It is also evident that by the 1950s, traditional Inuit life had been dramatically transformed. According to Edmund Carpenter, the Canadian Inuit had very much lost their belief in "magic" and shamanism by 1950 (although still maintaining some fear of witchcraft between ordinary people). Moreover, by the early twentieth century shamanism had been outlawed in many jurisdictions including Canada, Russia and the U.S. (Znamenski 2007, 60; Plotkin 1993, 205; Carpenter 1968).

In a recent study, Stephen and Suryani (2000) concluded that psychosis was distinct from shamanism based on their analysis of "folk healers" from the Indonesian island of Bali. The authors did, however, acknowledge that the comparison might not be entirely appropriate because their subjects offered contemporary "priestly functions," which likely required greater psychological organization than a traditional shaman. Leaving this aside, they found evidence for "initiatory madness" resembling schizophrenia in 18 of 108 (17 percent) magico-religious practitioners, and inexplicably concluded that this percentage indicated a weak association. In my view, a 17 percent prevalence is not trivial, and in fact suggests that the lifetime prevalence of schizophrenia in Balinese folk healers could be as high as seventeen times greater than that of the general population!

Bridging the Gap Between Schizophrenia and Shamanism

The observations in the Balinese folk healers study were remarkably similar to the raw data of the previously mentioned study of Japanese spiritualists, and yet the respective authors drew opposite conclusions. Where one researcher saw similarities, another magnified the differences. My view is that it is naive to expect shamanism to perfectly mirror schizophrenia, especially in light of the broad cultural gap between traditional societies and contemporary life. In fact, one does not even need to step outside of Western culture to find basic medical syndromes that, only a few generations ago, were perceived in dramatically different ways.

During the American Civil War a military physician by the name of Jacob Da Costa identified a medical syndrome consisting of periods of rapid pulse, shortness of breath (without exertion) and digestive complaints, occurring primarily in soldiers. The presumption was that these symptoms reflected some yet-undetected cardiac pathology, and the syndrome was referred to as "irritable heart." (It later became known as "soldier's heart" or Da Costa's syndrome) (Paul 1987). In hindsight, most these episodes, if not all of them, were what we would now call panic attacks.

The difference between nineteenth century American culture and that of the present is much smaller than the difference between hunting and gathering cultures and twentieth-century culture, and yet even over that comparatively small cultural gap we have a straightforward cluster of symptoms perceived in a radically different way. The physicians of the nineteenth century completely overlooked the anxiety component, even though it appears to be the essence of the problem (and if it weren't for the strong association between combat PTSD and panic disorder, this retrospective diagnosis may never have been identified). Accordingly, it would be unrealistic to expect anthropological accounts of life in foreign lands to include incontrovertible descriptions of schizophrenia.

As well as panic attacks, other not-so-subtle psychiatric symptoms such as posttraumatic symptoms and obsessive-compulsive behaviors have consistently slipped under the radar of historical observation. I would add minor psychotic thinking to this list because milder forms of insanity are probably more common than the extremes, yet descriptions of slightly disordered thinking are conspicuously absent from the anthropological literature.

Even for contemporary psychiatrists, identifying milder forms of psychosis can be touch-and-go, so I presume that it must have been almost impossible for nineteenth-century anthropologists, having no psychiatric experience and studying unfamiliar peoples. In fact, this prejudice — failing to identify insanity in alien cultures and attributing it instead to their customs — is so common that it was given an unceremonious name: Seligman's Error (Lucas and Barret 1995). Moreover, shamans are often secretive and reclusive, which would have further obscured matters.

On the other side of the ledger, it is certainly true that not all shamans were psychotic. Although I am convinced that the core shamanistic phenotype typically contains a psychotic component, human culture throws curve balls when it comes to complex behavior. Take, for example, the hominoid quality of leadership. Although there appears to be an evolutionarily-derived biological constituent (e.g., serotonin levels) that helps form social hierarchies, leadership behavior is also vulnerable to cultural forces. We all know people who seem to be constitutionally inadequate to be leaders but for one reason or another are thrust into leadership roles. Weston La Barre, one of the pioneering experts in shamanism, trenchantly emphasizes this very point (1970, 319):

> Some of the disputation over the shaman's mental state comes from diagnostic ineptitude, some from an undiscriminating and monolithic thinking about *the* shaman. Institutionally, shamanism is merely an identifiable social role. But as individuals, prophets and shamans run the full gamut from self-convinced and sincere psychotics to epileptics and suggestible hysterics, and from calculating psychopaths (more rare than commonly believed) to plodding naifs only following the cultural ropes.

One final point that deserves attention is the conspicuous lack of association between psychotic-like states and every other tribal social role. Positions such as warrior, chief, or craftsman are hardly ever linked to deranged behaviors in any persistent or substantive way. This lack of association cannot be easily dismissed. In contrast, there is a tendency for insanity, shamanism and religion to run together in traditional societies.

Evolutionary Functions Versus Shamanistic Functions

The essential argument of this book is that psychosis is a major constituent of shamanism, which in turn helps drive religion. The behaviors associated with shamanism are dictated by both instinctual and cultural factors. A lot more study and research will be required to better understand how evolution produced the genes behind shamanism. In trying to clarify this picture, one potential pitfall is that shamanistic functions must be separated from their evolutionary functions; some shamanistic activities may pack more of an evolutionary punch than others. By analogy, the primate hand can be used to either grasp or gesture; it is the need to grasp that has primarily shaped the anatomy of the hand, not the need to gesture.

Similarly, while shamanism encompasses diverse activities including spiritual healing, conducting rites of passage, divining, prophesying and giving war blessings (Radin 1937, 16), it cannot be assumed that every shamanistic function is evolutionarily advantageous. Determining which functions were the most evolutionarily important is beyond the scope of this book, but it is interesting to speculate. My own sense is that divining, prophesying and guiding rites of passage would be valuable to any community. Synchronizing a tribe could prove useful in many circumstances, particularly with respect to such potentially arbitrary matters as the timing of formal adulthood, the new direction of a hunting expedition or the decision to attack a neighboring tribe.

Another evolutionary purpose for shamanism could be "ritual healing" (McClenon 2002). It has been suggested the placebo functions of shamanism were reliant on hypnosis, and that hypnotic mental states eventually evolved into psychosis and religion (McClenon 2002). The fact that shamans all over the world spent a lot of time engaged in healing ceremonies supports this explanation. However, I have some reservations about the evolutionary importance of healing because it is possible that it is just an epiphenomenon. A faith healer's influence on survival, through instilling hope (the placebo effect) or using marginally effective medicines, may be overrated. The weightiness and uncertainty around sickness would have no doubt been a suitable fit for shamans: once the institution of shamanism was up and running (for other evolutionary reasons), it would have been natural to apply it towards the ambiguity of health problems (or, for that matter, the weather).

The influence of shamanistic behavior on inter-tribal warfare may be underrated (Peers 1994, 136; Murdock 1934, 45). Although war is relatively infrequent, the associated high death rates may produce a strong evolutionary impact. This dynamic could explain the pervasiveness of paranoid delusions in shamanism and its vestiges in schizophrenia. For example, among the Yanomamo, shamans routinely try to clairvoyantly cause sickness and kill people in other tribes (Ritchie 1996, 28). In Papua New Guinea, sorcerers are often called on to determine the cause of a man's death, and will often point the finger towards known adversaries or other tribes (Butcher 1964, 113). Among the Macushi people of Guyana, evil spirits almost always originate in neighboring tribes. According to Thurn, "Thus the feeling of mutual suspicion and hatred which has arisen naturally between the tribes is fostered and retained" (1883, 333). Thurn also described a man who hallucinated that warriors from a nearby tribe were lurking in the woods and subsequently tried to rally his tribe to counter-attack (which did not happen in this case) (p. 347).

The ubiquitous nature of paranoid ideation requires some evolutionary explanation; an attractive reason could be that it fosters militarism. Paranoid thoughts would increase vigilance and perhaps even encourage surprise attacks, thus giving the offensive tribe a huge upper hand. Referring to competition between businesses Andy Grove, the past CEO of Intel, aptly stated, "only the paranoid survive."

Conclusion

In describing shamans, John A. Grim (1983, 10) wrote, "Although shamans may not be the only religious figures in the community, they are unique because of the immediacy of their personal encounter with the spirit world and the vigor of the forces that they command." In traditional societies, people with a presumed biological propensity towards psychosis were almost always placed into the role of shaman. Prehistoric shamanism appears to have been universal and thus inevitable. These residual biologically ingrained behaviors persist, but are now so culturally out of place that they bear little resemblance to their native form.

– 9 –
On the Edge of Insanity

According to the shamanistic theory of schizophrenia, delusional ideas are particularly likely to turn into spiritual doctrine. Psychosis is therefore connected in some way to religion. Evidence for such a relationship can be seen in peculiar quasi-spiritual phenomena like paranormal beliefs, possession states and unconventional religious cults. These curious phenomena seem to straddle the domains of insanity and religion. Although it may be argued that any belief in a deity is delusional, conventional religious beliefs are accompanied by a set of actions which are considered normal behavior, as opposed to those of people immersed in psychotic thinking. The passive acceptance of religious knowledge is cognitively different than the generation of spiritual ideas (i.e., psychosis) — just like laughing at something funny is not the same as creating a joke.

Although the precise form of the psychosis-religion relationship is yet to be fully understood, schizophrenia may be at one end of a hypothesized continuum and mainstream religion at the opposite end, creating the fallacious impression that the two phenomena are only weakly related. I have previously tried to demonstrate that prehistoric religions exhibit more psychotic-like themes than contemporary monotheism. Psychosis in traditional cultures tends to be imbued with more conventional spirituality than psychosis in modern cultures, and so the gap between religion and psychosis appears smaller in hunting and gathering societies. However, even modernity cannot completely dissolve the links between religion and psychosis. Familiar social phenomena such as paranormal belief, demonic possession and cults are considered to be at the outskirts of both psychosis and religion.

This chapter will explore how the edges of insanity intersect with fringe religious beliefs. I will begin by comparing the hypothetical cognitive substructure of religious ideation to that of psychotic delusions. In the second section I will examine how religious content often makes its way into the delusions of schizophrenia patients. Third, I will explore three long-standing social phenomena whose cultural histories are unequivocally rooted in both religion and insanity: possession states, paranormal beliefs and hallucinogen use. Last, I will review how psychotic-like thoughts are frequently observed in the founders of new religious movements (i.e., cults).

The Cognitive Substructure of Religion Resembles Psychosis

We have previously examined how the major contemporary monotheistic religions assume a slightly different form than that of spirituality in traditional societies. Perhaps the greatest difference is that hunting and gathering societies are more concerned with ghosts, curses, possession and magical spirits while monotheistic religions emphasize moral guidelines and a spiritual bond with a single creator. There are, however, important shared properties. Prehistoric religions are similar to the major contemporary religions in their shared belief in non-physical beings, and in the presumption that specific people are especially likely to receive unique supernatural messages from gods or spirits (Boyer 2001).

Can further comparisons be drawn between prehistoric spirituality and contemporary religious faith? For example, is there an identifiable cognitive template underlying all religiosity? Until recently, a unifying theory outlining the fundamental cognitive structure of religious ideation eluded researchers. However, Pascal Boyer (2001) seems to have cracked the code, unveiling a universal cognitive mechanism behind all religious belief.

Boyer's theory begins with the presumption that the mind is structured to deal with five basic ontological categories: person, plant, animal, natural objects and man-made objects. In other words, the human brain is constitutionally set up to pigeonhole every object in the universe into one of only five categories. Further, each category is associated with a set of defining characteristics which clearly define which objects belong in that category. These divisions may or may not be mutually exclusive but the mind treats them as such. In other words, every object in the universe is labeled by a distinct category, like numbers on dice, not like colors on a spectrum. Perhaps that is why those rare hybrids such as a Venus flytrap (animal-plant), Hal in the story *2001: A Space Odyssey* (man-robot), the diabolical plant in *Little Shop of Horrors* (man-plant) and Pinocchio (boy-puppet) tend to catch our attention.

In Boyer's theory, a religious idea must incorporate a fact that specifically contradicts or violates at least one defining characteristic of its ontological category (Boyer and Ramble 2001). For example, a ghost may have the form of a person but it lacks one defining quality integral to its ontological category: physical matter. The resulting construct is a contradiction or violation of the defining terms of the corresponding ontological category. Zombies (active persons who are dead) are frequently referenced in Haitian Vodou, and can be classified as a religious idea because death is a direct contradiction of life. Another example is found in the spiritual folklore of the Aymara people, a community in the Andes, where a particular mountain is seen as a living creature, bleeding or feeding on sacrificial animals (Boyer 2001, 65).

I decided to randomly peruse some of my own anthropology books to see if Boyer's theory held up. Among the Dani people of the western New Guinea highlands (Heider 1970), deceased people can turn into ghosts and retain all of the regular anthropomorphic qualities, except that they can cause diseases and control the weather. There is also a Dani origin myth which tells that "men and birds once lived together in harmony, not realizing that they were different" (Heider 1970, 166). The origin myth of the Xingu tribes of Brazil involves a shell turning into a woman (Boas and Boas 1973, 53). Additionally, their spiritual folklore is full of animals (birds, jaguars and snakes) (Boas and Boas 1973, 175), trees and even stones that can talk (Boas and Boas 1973, 211). The Trobrianders of Papua New Guinea believe that sleeping witches can leave their bodies and fly around (Weiner 1998, 39). The Urubu (Ka'apor) people of Brazil believe shamans are capable of killing in magical (telepathic) ways (Huxley 1957, 198). And finally, the central Australian Aranda describe their sky god "as an enormous giant with a red skin, emu's feet and a retinue of wives with dogs' legs..." (Birket-Smith 1957, 23). The pattern is clear — ontological categories are being improperly mixed together in all of these magico-religious stories.

A number of stories from the Christian Bible fit Boyer's definition of a religious construct. Let's begin with God, the central spirit of the Bible, who is generally characterized as an omnipotent person lacking physical matter. The Immaculate Conception violates a very basic biological attribute of being human (i.e., requiring two human parents). In the first book of Samuel, an evil spirit is sent from God to Saul to terrorize him. Several miracles, like the parting of the Red Sea, represent mysterious action-at-a-distance which is normally considered impossible for human beings, but is characteristic of another ontological category, natural objects (e.g., gravity, wind, the movement of celestial bodies). Curses, which also represent action-at-a-distance, are found throughout the Bible, as well as in the folklore of perhaps every traditional society.

Now recall those ancient cave paintings and artifacts showing chimerical images that blended humans and animals together — believed to have been created by shamans. Initially, I only suggested that these hybrid images resembled contemporary shamanistic art, but by using Boyer's theory we can now move towards defining this artwork as religiously iconic. The oldest of these hybrid images, the lion-man statuette, which is a little over

30,000 years old, represents the first evidence of humankind's fascination with violations of ontological categories and could therefore provide clues as to the specific date when hominid brains acquired religion.

Notice that mere strangeness does not result in a religious idea. Bizarre or randomly put-together ideas will not seem religious unless they take on Boyer's specific cognitive form. I will make up some examples of strange ideas that do not qualify as spiritual themes: a forty-pound mosquito flying upside-down; a mute woman who hung two fish from her ears; every leaf in the forest turning blue; and a choir boy stretching his arm 1300 miles to smear asparagus-scented toothpaste on an alarm clock in Venezuela. These bizarre situations do not violate the defining characteristics of ontological categories. For example, the elasticity associated with stretching an arm 1300 miles may perhaps defy the laws of organic chemistry, but the concept itself does not contradict an essential quality of humanness. Similarly, although a forest suddenly turning blue may be unusual, changing the color of a leaf does not alter its essential qualities (i.e., leaves can naturally turn red or brown).

The distinction between magico-religious themes and more generic bizarre ideas also applies to psychosis. Contrary to popular belief, psychotic experiences are not a random assortment of non sequiturs, but have their own intrinsic structure. Although no consensus exists on their precise form, it appears to me that many common psychotic ideas resemble Boyer's religious forms. Examples include demonic influences, voices from God, telepathic communication and nihilistic delusions (such as feeling one is missing their vital organs).

In the 1930s, Kurt Schneider devised one of the most influential schemes used to distinguish schizophrenic psychosis from other types of insanity (Schneider 1959). Schneider divided schizophrenia symptoms into first-rank symptoms, which he felt were distinctive to schizophrenia, and second-rank symptoms, which could also be observed in other psychiatric conditions such as mania or dementia. Schneider proposed the following first-rank symptoms: audible thoughts, voices arguing, voices commenting on one's actions, the belief that bodily organs drive behavior, thought withdrawal or insertion enigmatically controlled by some external force, and thought broadcasting (Shorter 2005, 275). In contrast, second-rank symptoms (e.g., flight of ideas, nonspecific hallucinations and out-of-body experiences) were considered unhelpful in distinguishing schizophrenia from other psychotic conditions.

Notice how well Schneider's first-rank symptoms align with Boyer's magico-religious thoughts, as well as so many shamanistic experiences. For example, auditory hallucinations imply the presence of a human being without physical attributes — a clear ontological violation. Thought control and telepathic communication represent similar violations of the human condition. Out-of-body experience, which I would categorize as a religious idea, is the only Schneiderian symptom seemingly out of place.

Schneider's system caught on with psychiatrists and has been enormously influential in shaping the modern perception of schizophrenia; however, it is admittedly not the last word. Recent studies have doubted the exclusivity of Schneiderian symptoms in diagnosing schizophrenia (Hoenig 1984; Malik et al. 1990; Peralta and Cuesta 1999). Even Schneider acknowledged that his scheme was not perfect, and that some cases appearing to be schizophrenia lacked first-rank symptoms. Recent studies show that somewhere between 25 and 88 percent of patients with schizophrenia display first-rank symptoms (Nordgaard et al. 2008).

A number of studies have tried to identify fundamental patterns in the delusions of patients with either schizophrenia or schizoaffective disorders. Using principal component analysis, a recent study revealed three possible delusional subtypes (Kimhy et al. 2005):

i. Delusions of influence (being controlled, thought withdrawal, thought broadcasting, thought insertion and mind reading)
ii. Self-significance delusions (grandeur, reference, religious and guilt)
iii. Persecutory delusions.

Although this three-category grouping is not the final word on how psychosis is structured, I find it interesting that each subgroup contains common shamanistic themes. Notwithstanding a few blatant self-referential delusions (which tended to occur in the most disorganized patients) most psychotic experiences were either paranoid or compatible with magico-religious themes. To my knowledge, no one has yet explored psychotic thoughts and specifically compared them to Boyer's religious forms. In the next chapter, I present a pilot study that makes an initial probe into this question.

Spiritual Delusions in Schizophrenia

Although schizophrenia researchers have not yet performed analyses with Boyer's religious forms, a number of studies have tallied delusions associated with familiar religious themes (Kim et al. 1993; Yip 2003). This method is restrictive because it excludes magico-religious ideas that may not be associated with conventional religions; for example, a delusion of being controlled by an unseen device. Even so, most studies that have monitored generic religious delusions in schizophrenia show significant results (Gutierrez-Lobos et al. 2001; Kulhara et al. 2000) A small pilot study found noteworthy religious delusions or hallucinations associated with fifteen of twenty schizophrenic patients retrospectively sampled from a state hospital in Hawaii (Brewerton 1994). Similarly, 63 percent of 295 Lithuanian patients with schizophrenia experienced religious delusions (Rudaleviciene et al. 2008). Another study comparing symptoms of schizophrenia in African cultures found religious delusions in over 60 percent of 113 participants (Maslowski et al. 1998). A study that included both bipolar patients and schizophrenia patients showed religious or magical delusions in 56 percent of eighty patients sampled (Vlachos et al. 1997).

Another cluster of studies have demonstrated slightly lower proportions of classic religious delusions in schizophrenic patients: 24 percent of 193 British patients (Siddle et al. 2002), 31 percent of 123 Swiss and Quebecois patients (Mohr et al. 2010), 21 percent of 126 Austrian cases (Stompe et al. 1999) and 5 percent of 108 Pakistani patients (Stompe et al. 1999). Although these numbers appear comparatively small, it cannot be overlooked that magico-religious psychosis often resides in paranoid ideation. Of the aforementioned studies that also monitored paranoid delusions, 70–82 percent of schizophrenia patients were affected (Tateyama et al. 1998; Stompe et al. 1999).

Results showing that religious delusions are common in schizophrenia will not surprise anyone who has worked with psychotic patients. However, this predilection towards religiosity has always been chalked up to cultural influences: religion is an important aspect of life and therefore religious ideas permeate psychosis. However, that explanation may no longer be tenable. In my experience, twenty-first century urban patients rarely discuss spiritual issues — until they become psychotic. In fact, two studies that have examined this issue could not find any significant correlation between pre-psychotic religiosity and the frequency of religious delusions in schizophrenic patients (Rudaleviciene et al. 2008; Getz et al. 2001). These studies confirm that patients who have never shown an interest in spirituality have just as many religious delusions as those who are genuinely devout.

Possession States

The idea that demonic spirits or other entities can possess a person's soul has a very long history. Hunting and gathering societies throughout the world have believed that shamans can be invaded by spirits, which means possession experiences have likely been around for tens of thousands of years. Now in the twenty-first century, schizophrenia patients sometimes complain of being possessed and take antipsychotic medications to temper these experiences.

Prior to World War II, T. K. Oesterreich (1966) compiled an exhaustive list of possession accounts from antiquity through the Middle Ages and into the twentieth century, including accounts from traditional societies. Reading through Oesterreich's book it becomes abundantly clear that spirit possession has been a spiritual phenomenon in almost every culture — except modern Western nations, where such experiences are interpreted as insanity.

Oesterreich details dozens of possession accounts throughout the globe and across the centuries. One possession story originated in the chronicles of a nineteenth-century missionary living amongst the Kabyle people of Algeria. It involved a man described by the locals as "sick of the demon." He experienced "fits" that included self-mutilation, an inability to recognize familiar faces and gesturing wildly. Afterwards, he declared that an evil spirit possessed him (Oesterreich 1966, 133). Another vignette from the sixteenth century is eerily reminiscent of the 1970s film, *The Exorcist*: "The latter (a girl) was possessed by the demon who often threw her to the ground as if she had the falling sickness. Soon the demon began to speak with her mouth and said things inhuman and marvelous which may not be repeated..." (Oesterreich 1966, 9).

Not every case of possession, especially during the Middle Ages, would have been classical psychosis. Some cases were undoubtedly hysterical reactions. In fact, a number of studies imply that possession states frequently occur in hysterical individuals (Whitwell and Barker 1980; Somasundaram et al. 2008; Kua et al. 1993; Ward and Beaubrun 1981; Goff et al. 1991; Schreiber 2010). A 1950s Hong Kong study found "possession syndrome" in 66 patients and determined half to be hysterical and a quarter schizophrenic (Yap 1960).

Even so, schizophrenia (and perhaps bipolar disorder) would have been responsible for a substantial proportion of possession states in every epoch. In fact, studies that have specifically looked for demonic possession in schizophrenia have always found it in a substantial percentage of cases. For example, a study of 1029 Japanese inpatients, of which most were diagnosed with schizophrenia, revealed 21 percent had experienced delusions of possession (Iida 1989). Similarly, a U.S. sample of chronic psychotic outpatients showed delusions of possession in twenty-five out of sixty-one subjects (41 percent) (Pfeifer 1999). A larger study in Switzerland, involving devout Protestant outpatients, also showed high rates of belief in spirit possession (Pfeifer 1994). In fact, more than half of patients diagnosed with schizophrenia (53 percent) explained their hallucinations and delusions as the result of some type of occult force.

Paranormal Phenomena

Paranormal phenomena refers to a hodge-podge of experiences outside the domain of typical experience (and, some would also argue, outside the scope of scientific investigation). Some examples of paranormal phenomena are extra sensory perception (ESP), telepathy, precognition, astral projection, out-of-body experiences, reincarnation, channeling, telekinesis, ghosts, witchcraft, UFOs, and belief in cryptids (Bigfoot, the Loch Ness Monster). Most studies investigating these beliefs confirm that almost half the general population holds some paranormal beliefs (Irwin 1993; Goulding 2004). I suspect that for the most part these beliefs are soft — similar to how people can be a little superstitious. There are, of course, a small minority of people who espouse unshakable paranormal beliefs.

A number of psychological studies have tried to characterize devoted paranormal believers, but a comprehensive theory is lacking. It is not yet known whether paranormal belief is the result of a single trait, or of a variety of separate cognitive styles. Some studies have demonstrated schizotypal tendencies among persons with dedicated paranormal beliefs (Irwin 1993; Thalbourne 1994; Goulding 2005). Lawrence and Peters (2004) found that committed paranormal believers displayed more delusional ideation and made more deductive reasoning errors than individuals with weaker beliefs. Paranormal believers also tend to be religious, unmarried, hypnotically susceptible, imaginative and less reliant on analytical thinking (Irwin 1993; Lindeman and Aarnio 2006). Age and gender have shown little correlation (Irwin 1993).

Assessments of the psychological health of paranormal believers have yielded conflicting results (Irwin 1993; Goulding 2005). One problem is that most psychological studies have been designed so that they, in effect, exclude persons with schizophrenia by using college students as subjects, or relying on replies to newspaper advertisements. (Patients with schizophrenia are not apt to respond to such notices.) (King et al. 2007; Goulding 2005). One study suggested that although healthy paranormal believers may have schizotypal tendencies, they tend to experience less anhedonia and cognitive disorganization than schizophrenia patients (Goulding 2005). Notice that adding anhedonia and cognitive disorganization to schizotypal tendencies describes schizophrenia. It is also interesting that, in this study, 8 of 129 paranormal believers had apparently sought medical help for their paranormal beliefs at one time or another (and one was hospitalized briefly).

It may be counter-intuitive to monitor paranormal beliefs in schizophrenia because so many psychotic symptoms resemble paranormal beliefs, but one study did just that, finding paranormal beliefs in 38 percent of patients with schizophrenia (Kulhara et al. 2000). Moreover, 28 percent of patients had such beliefs prior to becoming ill; arguably a very similar rate to that of the general population. (Whether this figure is truly comparable depends on how passionately the schizophrenia group adhered to such beliefs.) Notice also how paranormal beliefs closely resemble Schneiderian first-rank symptoms of schizophrenia, such as thought broadcasting or sensations of motor control by external agents (Schneider 1959). Furthermore, paranormal ideas such as telekinesis, telepathy, precognition, reincarnation, channeling and belief in ghosts are almost identical to typical magico-religious shamanistic beliefs.

It is perhaps no coincidence that the word *paranormal* entered the English language in the early twentieth century, because in previous centuries such superstitious beliefs were commonplace. Historians seem to agree that intensely superstitious thinking was squeezed out by scientific rationalism, as well as established churches that saw such ideas as competition. (Conventional religions typically frown upon paranormal beliefs not affiliated to their own doctrines.)

In a series of scientific papers, neuroscientist Peter Brugger suggests that one way paranormal believers may be distinguished from the general population is by their greater tendency to see patterns in random configurations. A number of experiments have shown that believers in ESP (and other paranormal phenomena) tend to attribute significance to small random perturbations much more readily than non-believers (Brugger and Taylor 2003). In one study ESP adherents perceived significantly more "meaningful" images in random dot patterns than did non-believers (Brugger et al. 1993). Brugger points out that seeing meaning in unrelated items is not only seen in paranormal believers, but also in psychotic patients and creative individuals (Brugger 2001). Anecdotally, psychotic patients sometimes become unnecessarily concerned by innocent coincidences. They also have a greater tendency to perceive images within clouds, trees or smoke. Brugger concludes that this phenomenon may be related to right-hemispheric overactivity (Brugger and Taylor 2003).

Because Peter Brugger's observations resonated with me, I wanted to apply an objective test to find out whether schizophrenia patients would connect remotely associated stimuli better than the general population. Our lab constructed a straightforward task; we simply presented a series of images of varied blurriness and asked participants to guess the item in the picture. We presented the blurriest image first, followed by increasingly focused images, with more points awarded for earlier correct guesses. A total of twenty sets of images were presented (Figure 9.1) to twenty-eight schizophrenia patients and twenty-nine control subjects (education-, age- and gender-matched). Although schizophrenic patients did significantly worse on all the other cognitive tests (Trailmaking A and B, Color Stroop, and WAIS Digit Symbol), they did better than controls (38.2 correct versus

36.6) in the "blurry pictures" task. Although the difference fell short of statistical significance, it is nonetheless compelling because patients with chronic schizophrenia seldom outperform controls in cognitive testing.

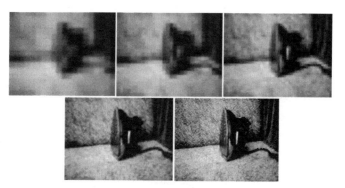

Figure 9.1: Each photograph was presented sequentially from blurriest to clearest. Participants attempted to guess the item in the photograph and received higher scores when requiring fewer hints.

Hallucinogens

In almost every region of the globe, hallucinogens were associated with traditional religion. Recall that the term *entheogen* is reserved for the category of mind-altering substances specifically affiliated with spiritual ceremonies. Entheogens generate transcendental feelings or hallucinatory experiences, especially when used in certain conducive environments (e.g., darkness, in the context of spiritual discourse). Shamans throughout the world incorporated entheogens in their arsenal of techniques to connect with the spirit world (Metzner 1998). In traditional societies, substance-induced hallucinations were generally interpreted as spirit visitations, which allowed for religious prognostication.

A variety of mushrooms were used in rituals by shamans on several continents (Furst 1976; Wasson et al. 1986). *Amanita muscaria*, also known as fly agaric, is a type of mushroom known to have been consumed by Chukchee shamans prior to religious rituals. *Amanita muscaria* contains the hallucinogens muscarine and muscimol (Michelot and Melendez-Howell 2003).

Southern Mexico has several species of mushrooms containing psilocybin, a substance that not only induces dream-like mental states (Hasler et al. 2004) but can also trigger hallucinations (Vollenweider et al. 1999). The local indigenous people, dating back to the time of the ancient Aztecs, routinely used these "magic mushrooms" in their divinatory rituals (de

Rios 1984, 138). This explains why the literal translation of the Mazatec word for this particular mushroom is "flesh of the god" (La Barre 1972, 276).

Recall from Chapter Seven Pahnke's Good Friday experiment from the early 1960s, which showed that psilocybin can cause transcendental feelings in religious people (Doblin 1991). Forty years later, a similarly designed study confirmed psilocybin's ability to generate mystical feelings in healthy subjects (Griffiths et al. 2006). Interestingly, another study monitored psilocybin's effects using entirely different psychological measures and concluded that it caused "schizophrenia-like psychosis" (Vollenweider et al. 1998).

Another hallucinogen, ayahuasca, was used by various Amazonian tribes (Reichel-Dolmatoff 1972). Ayahuasca is derived from the *Banisteriopsis caapi* vine, which is boiled to make a sacramental brew. The resulting potion contains a powerful hallucinogen whose active ingredient is harmine. Among the Kaxinawá people of Peru ayahuasca was taken not only by shamans but by all the adult males of the tribe (Kensinger 1973). The ensuing visions were believed to be influenced by spirits and therefore portentous. In the same Amazonian region, Jivaroan shamans also used ayahuasca to peer into the future (Kensinger 1973).

Peyote is a small cactus native to Mexico and the American Southwest (Furst 1976). It can be eaten raw and contains the psychedelic agent mescaline. Its use among indigenous peoples appears to have been described by sixteenth-century Spanish conquistadors. Among the Mazatec people and other nearby tribes, peyote was used in all-night religious ceremonies led by shamans (Lanternari 1963). Resulting psychedelic experiences were often interpreted as momentous spiritual encounters (Schultes 1972).

Several species of datura (also known as jimson weed) have been used both medicinally and in the context of spiritual rituals. A number of tribes throughout the American southwest and southern Columbian highlands were known to regularly consume this plant (Schultes 1972). The active ingredients are atropine, hyoscyamine and scopolamine, well-known psychedelic agents. Datura could be boiled into a tea which was frequently consumed by shamans.

The entheogens I have mentioned are only a few of the hallucinogens known to traditional societies. Many more psychotropic agents were used routinely by shamans, from undisclosed concoctions to familiar substances like tobacco, cannabis and *Salvia divinorum* (Albertson and Grubbs 2009; Dalgarno 2007). Some psychoactive substances, like tobacco, would have been used by almost every male in the tribe.

An interesting question is whether all this chemical indulgence throughout the millennia influenced brain evolution (Nesse and Berridge 1997; Sullivan and Hagen 2002). In other words, has the human brain undergone evolutionary changes as a result of being habitually soaked in psychotropic chemicals?[1] Would humans have been *more* constitutionally hyper-religious if it weren't for our ability to intensify religiosity through external means? Or, would shamans have been even more prone to psychosis if it weren't for the existence of hallucinogenic plants? The main point here is that the use of hallucinogens may have been so common that such bioactive chemicals became integral to the shamanistic phenotype (Harner 1973). When evolution was transforming shamanistic behaviors, it was doing so while the brain was usually awash in hallucinogenic substances. Therefore, the natural shamanistic phenotype may only exist in the context of regular hallucinogen use.

The enormous variety of psychotropic chemicals found throughout the world makes it difficult to discuss specific patterns of neurophysiological adaptation related to hallucinogen use in shamanism. Evolutionary adaptation requires some degree of environmental consistency that may not be present in shamanistic drug use. How can the brain adapt at the neuroreceptor level, to entheogens that vary in molecular form from one part of the globe to another? Ascribing one specific evolutionary function to disparate substances may be ill-advised, but neither can we completely dismiss potential evolutionary changes related to thousands of years of shamanistic exposure to mind-altering substances.

Drug Abuse in Schizophrenia

Interactions between the brain and psychotropic drugs are complex. Although habitual drug use is almost always destructive (especially in the context of modern life), lesser, more sporadic dosing may be harmless or even beneficial. The occasional stiff drink after a stressful day at work, for example, may be more medicinal than injurious. In fact, alcohol — up to two drinks per day — seems to have cardiovascular protective effects (Sacco et al. 1999). A study from Liverpool showed that small amounts of alcohol can actually improve dart-throwing accuracy — something to keep in mind if you are ever challenged to a game (Reilly and Scott 1993).

Schizophrenic patients abuse most of the common psychoactive substances, such as caffeine, nicotine, alcohol and marijuana, significantly above community norms (Degenhardt and Hall 2002; Selzer and Lieberman 1993; Regier et al. 1990; Kosten and Ziedonis 1997). The diversity of substances abused makes it difficult to propose a unifying mechanism, at

---

[1] I am always fascinated by how some health care professionals are taken aback by the ostensibly high levels of substance abuse in Western societies. Modern rates are comparatively low when one considers the preponderance of mind-altering substances used throughout hominid history.

a neurophysiological level, to explain the behavior. A desire to experience any altered state of mind might be the most parsimonious explanation of psychoactive substance use in both schizophrenia and shamanism.

A number of studies have shown that schizophrenia patients who regularly use cannabis demonstrate superior cognitive performance compared to schizophrenia patients who abstain (DeRosse et al. 2010; Loberg and Hugdahl 2009; Rodriguez-Sanchez et al. 2010; Yucel et al. 2010). Whether this means that cannabis actually improves cognitive symptoms, or alternatively that higher-functioning schizophrenia patients have a preference for marijuana, is not yet known. A similar result has been seen in bipolar disorder patients (Ringen et al. 2010). Interestingly, it appears that cannabidiol, a major constituent of marijuana, has potential antipsychotic effects (Roser et al. 2010; Zuardi et al. 2006). This stands in contrast to the other major ingredient of marijuana, tetrahydrocannabinol (THC), that can exacerbate psychosis. It is certainly possible that these contrasting ingredients play out differently in each person.

Nicotine is another chemical commonly used by people with schizophrenia. Smoking rates in schizophrenia are between 60 and 80 percent, which is two to three times community norms (Sagud et al. 2009; Winterer 2010). Although smoking has rightly been villainized due to its carcinogenic effects, its influence on the brain is a separate story. Nicotine possesses tranquilizing, stimulant, and appetite-suppressing effects, as well as potentially enhancing concentration in certain circumstances. There is growing evidence that schizophrenic patients preferentially benefit from the advantages of nicotine's effects on the brain. Some of the neurophysiological tasks which yield abnormal results in schizophrenia, such as auditory sensory (P50) gating, pre-pulse inhibition and eye-tracking, can become normalized in schizophrenic patients who use nicotine (Leonard et al. 2000). Similarly, in schizophrenia patients, nicotine seems to especially improve cognitive tasks which require a lot of attention (Winterer 2010).

Having enjoyed the occasional cigar, my subjective impression is that nicotine puts a damper on lively extraneous thoughts, which is consistent with the effects of a non-sedating tranquilizer. I am not sure how to better describe it — perhaps nicotine relieves the hum of faintly distracting ideas continually bouncing into consciousness. If my assessment is on the right track, it may perhaps explain why some extremely creative people — including Isaac Newton, Albert Einstein, Karl Marx, Sigmund Freud, Charles Darwin, Thomas Edison, and Gregor Mendel — have been especially fond of smoking (Gately 2001).

The predilection towards substance abuse in schizophrenia has a passing resemblance to a shaman's indulgence in entheogenic chemicals. This parallel may simply be a coincidence; however, it could also represent some sort of vestigial evolutionary function related to shamanism. Historical clues about shamanistic entheogen use can not only help us understand substance abuse in schizophrenia, but perhaps also guide us in the quest for better therapeutic drugs.

New Religious Movements

New religions are continually being born. The anthropologist and pioneering expert in aboriginal religions Weston La Barre once wrote, "Every religion, in historical fact, began in one man's 'revelation' — his dream or fugue or ecstatic trance" (La Barre 1972, 265). This important insight emphasizes that, although religious institutions may evolve with the input of many, their genesis is typically inside a single head. I would add that almost all of these "revelations" are rooted in psychotic or quasi-psychotic thought processes. Not only prehistoric belief systems, but contemporary religions too, are often spearheaded by shaman-like people teetering on the edge of insanity.

Religions associated with larger civilizations are clearly more formal than their pre-literate counterparts. Although eccentric people may be able to conduct shamanistic rituals in small indigenous tribes, they generally cannot function within the ossified bureaucracy of modern religious institutions. Combining quasi-delusional magico-religious ideas with charismatic leadership often produces a spiritual redemptive quest that appeals to marginalized people. The ensuing societal fracture results in an entirely new religious sect, a process that has repeated itself many times throughout civilized history. The evolutionary psychiatrists Anthony Stevens and John Price (2000) have written expertly about the potent combination of schizotypal personality traits and charismatic leadership.

The new religious movements of past centuries usually involved marginalized segments of society, and as such may not perfectly mimic how an entire tribe would react to a quasi-psychotic shaman. Nonetheless, stories describing the inception and propagation of new religious movements may be a window to how the first spiritual myths were formed by prehistoric peoples thousands of years ago.

Some of the oldest records describing new religious movements are found in the Bible. In their beginnings, Judaism and Christianity resembled many of the new-fangled twentieth century cults. Stories of disenchanted peoples led by prophets claiming communication with spiritual entities have reoccurred over the last several thousand years (Stark 1999). Any inference about the psychological state of biblical characters — many who were probably not real people — is highly speculative. Nonetheless, biblical prophets generally received the word of God through visions and dreams — just like traditional shamans. Moses talked to God "mouth to mouth" while Jesus claimed to be the son of God — testimonies that nowadays would certainly result in psychiatric admission.

In contrast to Biblical figures, some authentic biographical information is known about the life of the Islamic prophet Mohammed. Living in the seventh century, Mohammed sometimes meditated alone for days in a cave at the foot of Mount Hira (Kurtz 1986). In his early forties, Mohammed began to be distressed by depression, hyper-religiosity, visions and auditory hallucinations. At first, Mohammed was confused by these unusual experiences. However, his spouse, instead of being skeptical, apparently fed into his hyper-religious beliefs (Kurtz 1986). Mohammed became increasingly convinced that the angel Gabriel was communicating with him, and that he was a prophet of God (Freemon 1976).[2]

One of the earliest cult leaders for whom we have substantial biographical information is George Fox (Ingle 1994). During the seventeenth century Fox founded the Religious Society of Friends, known as the Quakers, based on a number of religious visions and divine revelations. He experienced auditory hallucinations that, to him, were communication from God. Biographical accounts described him as anxiety-prone and behaviorally erratic. He had periods when he preferred solitude, while at other times he could be gregarious. Fox is thought to have had at least one bout of psychogenic blindness. Another time, he had a telepathic premonition concerning the persecution of his fellow Quakers. It was said of Fox by Lord Macaulay, "He had an intellect in the most unhappy of all states, that is to say, too much disordered for liberty, and not sufficiently disordered for Bedlam" (Middleton 1931). This insight seems to apply to almost every cult leader.

Mother Ann Lee was the founder of the Shakers (United Society of Believers in Christ's Second Appearing), a late 1700s religious sect. Their religious services tended to be informal, with singing, dancing, jumping and ecstatic shaking (Whitworth 1975). It has been suggested that these hysterical antics may have represented a form of spirit possession (Garrett 1935). Mother Lee claimed to have received revelations directly from God, and that she was the resurrected female counterpart of Jesus Christ (Evans 1858). She was arrested multiple times for boisterous and oppositional behavior; this behavior was interspersed with bouts of depression. Before coming to America, Ann Lee was admitted to a Manchester asylum in 1770 (Francis 2000). Ann Lee lost four of her children during childbirth (or soon after); the emotional burden probably contributed to her emergent psychosis. It also provides a possible explanation for her view that sexual

---

[2] Mohammed's father died before he was born and his mother died when he was six years old — the latter event can increase the probability of depression and psychosis in later life.

relations were wicked (Ingoldsby 2006). Shakerism is now almost extinct, in part due to this unusual sexual prohibition.

In contrast, the Seventh-day Adventist Church, founded in the nineteenth century, is a prosperous American religion with millions of adherents. One of its founders, Ellen White, reported almost-daily visions at certain periods of her life (Kurtz 1986, 267). Initially her family chalked these experiences up to hallucination, but later considered them divine. Ellen White claimed to talk to angels regularly, and on at least one occasion to have ascended to Heaven and conversed with Jesus.

Mary Baker Eddy, the founder of Christian Science, also heard voices in childhood (Ludwig 1995, 132). Other notable theologians who experienced religious hallucinations were St. Augustine, Thomas Aquinas, Martin Luther, Ignatius of Loyola, St. Teresa of Avila and Emanuel Swedenborg. However, none of these figures appear to have been grossly impaired by other psychiatric symptoms (Buckley 1981; Cangas et al. 2008; Folsom 1880; Johnson 1994; Pierre 2010; Storr 1996).

Another popular American religion with millions of followers is the Church of Jesus Christ of Latter-Day Saints, whose devotees are known as Mormons. Their founder, Joseph Smith, trumpeted a number of spiritual revelations that could be interpreted as psychotic experiences. He claimed that an angel by the name of Moroni appeared to him and told him that God had work for him to do (Kurtz 1986, 237). Joseph Smith's parents were superstitious people and, like Mohammed's wife, they encouraged Joseph to do what the angel requested (Riley 1903).[3] Beginning at age fourteen, Smith claimed to have had a number of visions, including seeing a divine "pillar of light" and a person suspended in air over his bed (Riley 1903, 53). He maintained that the angel Moroni had directed him to certain golden plates — the source of the Book of Mormon. He also declared that God resided near the planet Kolob.[4]

There have been accusations that Joseph Smith was a charlatan. Smith's delusions were in fact very ordinary compared to those of most shamans or schizophrenics. Unlike the majority of psychotic experiences, Smith's delusions were not decisively persecutory, nor did any of them satisfactorily fulfill Boyer's definition of a religious idea. Therefore, in my mind, the door is open to the allegation of charlatanism. (Although Smith may have had bipolar disorder, which tends to be associated with less-bizarre delusions.) For leaders of most new religious movements, however, the

---

[3] Joseph Smith's parents believed in a variety of supernatural phenomena including witchcraft and divination (Riley 1903, 70). His father saw visions (p. 23), and his mother heard voices (p. 22).

[4] "Although many people are aware that one of Joseph Smith's brightest and most appealing sons, David Hyrum, tragically lapsed into insanity and spent the last years of his life in a mental institution, few realize at least six other male descendants of the Mormon prophet also suffered from psychological disorders, including manic-depression" (Foster 1993, 10).

conviction of belief is typically deep and sincere. A number of cults, for example, have come to their suicidal end due to the genuine nihilistic delusions of their leaders. Some notable examples include:
- Luc Jouret (a physician) and Joseph Di Mambro, founders of the Order of the Solar Temple. They were among fifty-three who apparently killed themselves in Canada and Switzerland in 1994. Jouret apparently believed that their deaths would "secure an immediate magical contact with the spirits of fifty-four Templars burned at the stake in the fourteenth century" (Lewis 2001).
- Joseph Kibweteere, an excommunicated Roman Catholic priest, who founded the Movement for the Restoration of the Ten Commandments of God. In 2000, he heard a conversation between the Virgin Mary and Jesus intimating an imminent apocalypse. Kibweteere was among over 500 people who died. (There were claims that some people were coerced to kill themselves or even murdered.) (Lewis 2001)
- In 1997, thirty-nine people were found dead in an upscale mansion in San Diego, apparent participants in a mass suicide. The deceased had been part of an eccentric "UFO cult" called Heaven's Gate that believed they would all spiritually ascend to a UFO after their deaths. Marshall Applewhite and Bonnie Nettles, who called themselves Bo and Peep, founded this bizarre religion twenty years earlier. It appears they believed their bodies were occupied by extraterrestrials, akin to spirit possession (Lewis 2001).

Perhaps the most "mainstream" of the cult-like religions is the Church of Scientology (Bainbridge and Stark 1980; Miller 1998; Wakefield 1991). In the 1950s L. Ron Hubbard founded Scientology, which evolved from an over-reaching, pseudo-scientific psychological treatment program called Dianetics. Hubbard may have had bipolar disorder. He was an imaginative and prolific science fiction writer. According to biographical accounts, he could be temperamental, erratic, paranoid, phobic, depressed and grandiose. His most unusual delusion was that he had visited heaven on another planet, forty-three trillion years ago; and then again forty-two trillion years ago.

The Greek mathematician Pythagoras also established a pseudo-scientific religious cult in Crotone, a Greek colony in Southern Italy, 2500 years earlier. While portraying himself as a sort of demigod, Pythagoras advanced a number of metaphysical-style spiritual teachings (Riedweg 2005). Pythagoras came up with the idea that the universe contained superior beings, devoid of physical matter, that guided humans on earth. He also believed in reincarnation, even suggesting that he had once been a plant (Pickover 2009). It is believed that Pythagoras heard voices, and once withdrew into a nearby cave (as shamans do) for meditative solitude (Leudar and Thomas 2008; Pickover 2009).

A number of political movements and military campaigns have been raised to the status of cults by leaders with shamanistic personalities (Allen 1975; Ireland 1883; d'Orsi and Tinuper 2006; Richey 2000). The story of Joan of Arc (1412–1431) is one of those fascinating turns of history. Joan of Arc claimed that since age thirteen she had experienced regular visions and heard the voices of saints and angels. During the Hundred Years War, a desperate and discredited French army with seemingly few rational options allowed this illiterate peasant girl to lead their war effort against the English. As a teenager, Joan of Arc gained the attention of the French elite through her unique charisma — a potent blend of energy, intelligence, temerity and epical passion. She showed hints of divine proclivities and her spiritual fervor emboldened French troops to a number of victories before she was eventually captured, put on trial and burned at the stake.

Louis Riel was another devout and impassioned leader who became Canada's most famous rebel as a champion of Métis rights. (The Métis are people of mixed European and Native ancestry.) During the late nineteenth century, he led two rebellions in the Prairie provinces (Manitoba and Saskatchewan) against the newly established Canadian government. Riel experienced periods of uncontrollable crying interspersed with erratic outbursts, including tearing off all his clothes in public (Perr 1992). He had spent two years in a Quebec mental asylum with psychotic symptoms and acquired a diagnosis of "delusions of grandeur" (Patterson and Lee 1992). He almost certainly had bipolar disorder. A deeply religious man, Riel genuinely believed he was a prophet, which caused the Catholic Church to distance itself from his movement. Riel was eventually captured, put on trial for treason and executed.

Adolf Hitler, one of the most enigmatic figures in history, unfathomably managed to lead an established European nation down a bizarre self-destructive path. According to Stevens and Price (2000), Hitler was a charismatic shamanistic-style leader with a psychiatric diagnosis of schizotypal personality disorder (as well as narcissistic, antisocial, borderline and hysterical personality traits). Germany's economic misery of the 1930s made it ripe to be led by a fanatical leader who could "empower it with a vision of salvation." In Hitler's biography Stevens and Price point to a number of odd mystical beliefs, excessive suspiciousness, incoherent tirades and possible auditory hallucinations (and visions). Moreover, Hitler had a predilection towards superstitious thinking and may even have had some interest in the occult. By all indications, the Fuhrer believed that only he could fulfill Germany's providential destiny of greatness (Sickinger 2000). A posthumous study asked a group of historians to fill out personality questionnaires on Hitler which resulted in high scores on the schizophrenia scale. (Interestingly, this study also suggested Hitler may have had PTSD, presumably from World War I.) (Coolidge et al. 2007) Other evidence consistent with a diagnosis of schizotypal personality disorder relates to Hitler's maternal aunt, who apparently had schizophrenia (Stevens and Price 2000).

Hitler's mesmerizing presence prompted Carl Jung to compare him to a shaman (Sickinger 2000); and under Hitler, the Nazi party transformed itself into perhaps the largest cult the world has ever seen (Stevens and Price 2000, 96):

> ...Nazism had its Messiah, its Holy Book (*Mein Kampf*), its cross (the swastika), its religious processions (the Nuremberg Rally), its rituals (the Beer Hall Putsch Remembrance Parade), its anointed elite (the SS), its hymn (the Horst Wessel Lied), excommunication and death for heretics (the concentration camps), the devils (the Jews), its millennial promise (the Thousand Year Reich), and its Promised Land (the East).

The spellbinding power of magico-religious shamanistic leadership also has a lighter side, seen in modern popular music. Rock musicians often borrow elements of shamanism to enhance their mystique, whether it's being mysterious like Prince, or sacrificially burning one's guitar like Jimi Hendrix. However, the most shamanistic elements of pop music are contained in its lyrics. In the 1960s, Bob Dylan was among the first pop songwriters to utilize a stream-of-consciousness lyrical style which mimics the sort of loose associations often seen in psychotic ramblings. A typical Dylan line like, "The dead will arise and burst out of your clothes..." is both magico-religious and quasi-psychotic.

John Lennon wrote "Across the Universe," a song that was among his Beatle favorites, late one night in 1967. Part of the song's emotive power originates in the magico-religious texture of lines like, "Sounds of laughter, shades of earth are ringing through my open views," or "Images of broken light which dance before me like a million eyes." Another song, "I Am The Walrus," apparently written during one of Lennon's acid trips, contains the quizzical lyric, "I am he as you are he as you are me and we are all together." This shamanistic-style line is similar to another, "I know that everything that you are, you are through me, and everything that I am, I am through you alone!" — spoken by Adolf Hitler (Stevens and Price 2000, 97). (Notice that Hitler's version is completely self-referential while Lennon's is more pro-social, as he adds that "we are all together.")[5]

---

[5] Although there did not appear to have been any notable psychiatric disturbances among the Beatles, John Lennon's mother may have had bipolar II disorder. She was inexplicably pushed aside by her family as the primary caregiver of her children. According to John

Paul McCartney has the reputation of being less philosophical than John Lennon; however, his lyrics too evinced powerful magico-religious themes. The song "Let It Be," for example, contains the following lines: "Mother Mary comes to me ... she is standing right in front of me" and "When the night is cloudy, there is still a light that shines on me."

The lyrical development behind the Beatles most commercially successful recording, "Hey Jude," is another exceptional demonstration of how the injection of magico-religious material bolsters the inspirational dynamism of a song. The song began as "Hey Jules," written to encourage Julian Lennon through his parents' painful divorce. McCartney said, "I changed it to 'Jude' because I thought that sounded a bit better" (Miles 1997, 465). I suspect McCartney intuitively recognized that the new line elevated the song in a spiritual sense. The song contains a highly repetitive four-minute tag ("naa-na-na nanana-naa") resembling a religious chant or mantra, as well as (at the risk of making too many shamanistic parallels) McCartney singing some ad lib gibberish akin to speaking in tongues. The line, "The movement is on your shoulder" was going to be changed because McCartney felt that it sounded "crummy," as well as making absolutely no sense. John Lennon, however, thought it was the best line in the song and convinced him to keep it. In his authorized biography, McCartney says that he regularly receives letters from "religious groups and cults," telling him how deeply inspirational those words are to them (Miles 1997).

All over the world new religious movements, often led by persons with schizotypal personality traits, spring up within large stratified civilizations; the examples I have provided only scratch the surface. Many other well-known cult leaders were probably delusional, including Jim Jones, David Koresh, Charles Manson and Shoko Asahara (who was responsible for the 1995 sarin gas attack on the Tokyo subway) (Lindholm 1990; Storr 1996). Admittedly, this list of biographical vignettes is not strictly "scientific" and it would be legitimate to consider such evidence anecdotal. However, given the relative scarcity of schizotypal personality disorder in the general population, there certainly appears to be a disproportionate number of such instances. In any case, I believe that it is possible to make such a hypothesis testable. For example, it is conceivable that one could someday recruit a dozen or so cult leaders and test them with a battery

---

Lennon's sister, their mother was "unusual, unpredictable and talented"; and "always laughing and joking" (p. 11) She was also known to "ride bicycles with her feet in the air, run races, skip, juggle with three balls, do head-stands, hand-stands, the upside-down crab and cartwheels," as well as "draw, paint, sing, dance and play instruments" (p. 63) (Baird 2008).

of biological correlates of schizophrenia, such as eye-tracking deficits or EEG markers.

Conclusion

The main point of this chapter was to explore those edges of insanity that intersect with fringe religious beliefs. We began by examining similarities between Boyer's cognitive substructure of religion and typical psychotic thoughts. Schizophrenic delusions are not only religious in form, but also in content, frequently containing familiar religious themes. In addition, there are a number of cultural phenomena such as possession states and paranormal encounters that undoubtedly incorporate elements of both spirituality and psychosis. The ancient use of hallucinogens — so common in traditional societies — also joins religion to psychosis.

Last, we followed the histories of new religious movements and discovered that they are almost always spearheaded by single individuals. Such cult leaders tend to be hyper-religious and mildly psychotic.

The amount of crosstalk between psychosis and religion requires explanation. Humankind's genetic endowment related to shamanism is perhaps the most economical explanation of the connection between psychosis and religion.

– 10 –
Contemporary Delusions and Hallucinations

A few years after publishing our original paper on shamanism, I began to wonder whether there were any clues within schizophrenic delusions or hallucinations that could shed light on the shamanistic theory. Studying delusions and hallucinations is notoriously difficult and unreliable. First, patients often conceal the details of their psychotic ideas because most have enough insight to recognize that they will be considered peculiar when they express certain kinds of ideas. Second, how does one even begin to categorize such wide-ranging, fanciful stories? Even a simple delusion about, say, being pursued by a demon can be classified in a variety of different ways: paranoid, religious or even grandiose — what makes you so important that a demon would be interested in you?

If the shamanistic theory of schizophrenia is on the right track, shamanistic psychoses should translate into an evolutionary advantage and because the theory assumes group selection is operating, those benefits should be bestowed on the entire tribe. We have already learned from anthropological accounts that many shamanistic behaviors do seem to propagate magico-religious ideas, as well as warn of danger from other tribes. Fostering religion and boosting tribal vigilance both qualify as evolutionarily advantageous group-selected phenotypic traits. If schizophrenia and shamanism are related, are remnants of these shamanistic themes evident in contemporary psychosis? In other words, could magico-religious themes and out-group paranoia be so genetically ingrained that these types of delusions might be assessed methodically in twenty-first century schizophrenia patients?

A Pilot Study

With these questions in mind, Jeff Reiss and I designed a study to directly assess the content of delusions and hallucinations in present-day schizophrenia patients. Actually, it wasn't much of a design — we simply asked patients about their delusions and jotted them down. We exercised the typical requirements for a psychiatric study: patients had to have a DSM-IV diagnosis of schizophrenia or schizoaffective disorder, they had to be between the ages of 18 and 65, and we excluded anyone with a history of head injury, seizure disorder or dementia. During the interview we took notes because audio recording devices can be intimidating and inhibit free expression.

Each patient was asked six questions about possible psychotic experiences. (As well, with the patient's permission, we reviewed their medical chart.) The first four questions were fairly general; the last two specifically asked about spiritual themes. I would therefore consider the last two responses to be prompted. I don't believe that prompting results in patients making up stories related to the content of the question, but it is a possibility and therefore we have indicated prompted responses by the letter P in the results section. (Patient 11, however, showed such disordered thinking that I suspected my questions may have prompted his responses.)

The questions were as follows:
i. Have you ever had a belief that others found unusual or that others did not believe?
ii. Have you ever heard voices others did not hear or seen something others did not see?
iii. Have you ever felt that others were against you, or that you were in danger, or had paranoid feelings?
iv. Have you ever had an unusual experience?
v. Have you ever had a unique religious or spiritual experience? For example, about God or the devil? (Prompted question.)
vi. Have you ever had a mystical or paranormal experience such as a premonition, witchcraft, magic, the occult, ESP, UFOs or an out-of-body experience? (Prompted question.)

Any affirmative answers were further clarified in order to rule out drug-induced psychosis, as well as to examine the intensity of each hallucination or delusional belief.

We also asked whether the object of paranoid ideation was familiar. We expected that paranoid delusions about unfamiliar people would be more evolutionarily advantageous than those concerning people you know, so we made a distinction between out-group and in-group paranoid delusions; we expected out-group paranoid delusions to be more common. In our study, out-group paranoid delusions were defined as situations where patients had either seldom contacted or never encountered the objects of their delusion. In contrast, any contact with the object of a paranoid delusion within one month defined it as an in-group paranoid delusion. Paranoid delusions about close friends or relatives, even with sporadic contact, were also considered in-group.

During the course of the study we identified a few small problems. For example, the first eight patients signed a consent form that included the original title of our study, "Exploring Spiritual Ideation in Schizophrenia." We realized that a title about spirituality could inject bias and so we subsequently used a more generic title, "Investigating Delusions and Hallucinations in Schizophrenia." Also, it was only after we had begun the study that it dawned on us that a control question should be asked in order to test whether "prompting" was a significant issue, so we asked the last twenty-two of twenty-six subjects if they had ever had any concerns or hallucinations about "green monsters." Of the twenty-two patients asked, not one reported a delusion or hallucination about a green monster.

Here is a breakdown of the results:

- Nineteen of twenty-six (73 percent) patients with schizophrenia or schizoaffective disorder described paranoid delusions. (Notice that this is consistent with similar studies, which usually report figures in 70–80 percent range.)
- Of the nineteen paranoid delusions, seventeen demonstrated at least one out-group paranoid delusion.
- Seventeen of twenty-six (65 percent) had delusions containing traditional religious content. Fourteen of twenty-six (54 percent) demonstrated unprompted evidence of religious content in their delusions, which is comparable to figures in similar studies.
- Twenty-four of twenty-six (92 percent) had magico-religious delusions, and twenty-three of twenty-six (88 percent) had unprompted evidence of magico-religious ideation.
- Remarkably, twenty-five of twenty-six (96 percent) had either magico-religious or out-group paranoid delusions.

The delusions of twenty-six outpatients with schizophrenia or schizoaffective disorder from a downtown hospital in Winnipeg, Canada between July 2007 and June 2009 are presented in the next section. One patient withdrew consent during the question period so those records are absent. Each entry represents our arbitrary attempt to separate each delusion and hallucination.

Some patients were actively delusional while others were describing prior episodes of psychosis. It is intriguing that almost all patients intuitively understood what I was getting at by my line of questioning. In other words, they generally knew that they had experienced a partitioned set of beliefs qualitatively different from normal experience.

Delusions of Modern Patients with Schizophrenia

Participant 1: 35-year-old male

- He acknowledged that God had in the past communicated with him directly.
- He figures that people around him sometimes speak (literally) in a certain code that he does not understand.
- He acknowledged periods of being suspicious that his girlfriend was cheating on him. The participant said that, in hindsight, he was likely paranoid, because the suspiciousness was not based on any tangible evidence.
- "I think I once saw a UFO."
- (P) "I think God talks to me."
- (P) "God gets mad at me."
- (P) The participant acknowledged having dreamed about some events that actually occurred "two weeks later."
- (P) In the past, "I took an interest in astral projection but now I think it's BS."

Deduction: religious delusion, paranoid delusion (out-group and in-group)

Participant 2: 40-year-old male

- In the past, "I thought I saw God."
- In the past, "I thought I killed the devil."
- In the past, "Zeus was making me feel pain."
- Ten years ago, "I thought the police were after me."
- (P) In the past, "I had an out-of-body experience ... I was floating."
- (P) In the past, "I thought there were UFOs outside my window."

Deduction: Religious delusion, paranoid delusion (out-group only)

Participant 3: 34-year-old male

- In the past, "I thought people were giving me AIDS (in tainted blood)."
- In the past, "I thought professional killers were trying to kill me."
- In the past, "I felt people were watching me through the TV."
- (P) In the past, "I felt God was talking to me (in a movie theatre)."
- (P) The participant said that sometimes he feels that things will happen before they actually do "...like the bus comes." "I felt John [a friend] was going to die before he did."

Deduction: religious delusion, paranoid delusion (out-group only)

Participant 4: 38-year-old male

- The participant once phoned a friend at 3 a.m. because he felt that "Canada was going to go bankrupt [that night]."
- The participant previously believed "mother is spying on me... she's everywhere." He believed his mother could control or influence his mind.
- (P) In the past, he thought that an ex-girlfriend was talking to him through a medium or through the voices of other people.
- (P) The participant said that he once saw himself outside his body after running into a goalpost. I asked for clarification suggesting that he may have perhaps just felt disoriented but the participant was emphatic, "yes, I was actually outside my body."
- (P) The participant said that he experienced regular premonitions, "I think bad things will happen to people I know." Moreover, he believed that these "bad things" actually do happen after he thinks about them.

Deduction: magico-religious delusion, paranoid delusion (in-group only)

Participant 5: 33-year-old male

- In the past, "I thought I could be the reincarnation of Jesus."
- In the past, the participant thought, "aliens [had] invaded the planet."
- In the past, he believed that "people were out to kill me." He didn't know who these people were — perhaps, street gangs.
- (P) The participant said, "I had a dream of seeing the devil." When I asked for clarification, suggesting that the participant was only dreaming, he replied, "I felt it was significant."
- (P) "I sometimes feel I can predict things... I think there's something to that."

Deduction: religious delusion, paranoid delusion (out-group only)

Participant 6: 34-year-old male

- In the past, the participant believed, "Two demons [were] speaking to me, put there by the devil, just to torture me."
- In the past, he believed, "The devil put a chip in my ear so they can talk."
- The participant said that people on the street can sometimes look menacing. He sometimes fears that "they could have shotguns under their trench coats."

Deduction: religious delusion, paranoid delusion (out-group only)

Participant 7: 47-year-old female

- In the past, the participant believed, "My mind was trapped in a book. I was trying to get my mind out of this book."
- (P) "I saw a demon and a monster on the street."
- (P) In the past, she believed, "demons [were] out to get [her]."
- (P) The participant once had a premonition that her brother was going to die.

Deduction: magico-religious delusion, paranoid delusion (out-group only)

Participant 8: 64-year-old male

- In the past, "I thought I was dead."
- (P) In the past, "The devil was speaking to me."

Deduction: Religious delusion

Participant 9: 28-year-old male

The participant denied any psychosis during interview, however, the chart revealed:

- Delusional thoughts that others were doing some sort of evil to him.
- The participant had said that he was chosen by the "makers" to experience hallucinations.
- The participant suspected that his roommate was stealing from him (but physicians believed it was delusional).
- The participant believed that he was fated to marry a specific woman.

Deduction: magico-religious delusion, paranoid delusion (out-group and in-group)

Participant 10: 45-year-old male

- The participant sometimes heard an "echo voice" which represented a visitation from dead relatives.
- He claimed to be a "time traveler." He said, "God, in space, gave me the ability [to time travel]." He added, "I died in 1999 and came back from a star in 2001." He also said, "I was sent through time from the past, 1000 years ago." He further added, "Queen Elizabeth died in the future."
- (P) The participant said he had a premonition about lottery tickets, "I won but lost the ticket."
- (P) The participant said, "I was at the last supper with Jesus." "I am a disciple of God ... I'm a very religious person."

- (P) When asked about out-of-body experiences, the patient replied in the affirmative. He said that he was "at the edge of the universe with God."

Deduction: religious delusion

Participant 11: 40-year-old male

- "I'm the man from beside God."
- "I told these girls, I'm going to go visit God ... and then Zion."
- (P) "I've predicted rainbows."
- (P) "I'm a prophet."
- (P) The participant said that he had, "One out-of-body experience ... looking at myself in bed."
- (P) "The devil, I beat the shit out of him." (The participant's speech was disorganized and I suspect he may have been prompted by some of my questions.)

Deduction: Religious delusion

Participant 12: 49-year-old female

- In the past, "I used to believe in aliens."
- In the past, "I believed I can influence Cory Hart [famous Canadian singer]."
- In the past, "I believed I had two mothers ... one was abducted by aliens."
- (P) In the past, "I believed Mary would tell God my prayers."
- (P) In the past, "I believed Mary appeared as me to [my boyfriend] X"
- (P) In the past, "A guardian angel came ... I don't understand that part."

Deduction: religious delusion, paranoid delusion (out-group only)

Participant 13: 51-year-old male

- The participant said, "The spiritual opening has eyes."
- The participant said that "people know what I'm thinking; they know when I'm in the show." (He didn't know these people.)
- The participant said that he had previously heard "voices on the radio." He also said, "He uses another human form to come to my place," but even with clarification, I could not understand his reference.

- (P) The participant had practiced Buddhism and recalls having felt "an energy coming into me." The energy had healing properties, "My sore throat was better without medicine."
- (P) A "grandmaster got into my body through my anus." I then asked, "How did he get into your body?" He divided his spiritual body: "He can divide into many bodies."
- (P) The participant also talked vaguely about his previous lives.

Deduction: religious delusion

Participant 14: 36-year-old female

- The participant said that people could somehow "transform" inside her. For example, she may "gain twenty pounds" because someone else has become "a part of me."
- The chart had previously quoted the participant, "Someone is pushing nerves in my mouth" and "People changed my teeth."
- Referring to the participant, the chart notes indicated, "People spying on her, controlling her."

Deduction: Probable magico-religious delusion, paranoid delusion (out-group only)

Participant 15: 44-year-old male

- The participant felt that his parents wanted to harm him in some way.
- The participant said that last week he believed he was Jesus Christ.
- The participant believed he has communicated his thoughts telepathically to other people.
- (P) The participant was once involved in a group prayer where he said he felt a "tingle" and thought it may have been a signal from God.
- (P) The participant once had a premonition that, "I would be a father and have a daughter."

Deduction: religious delusion, paranoid delusion (in-group only)

Participant 16: 59-year-old male

- The participant had previously believed that India or China had declared war on Canada.
- The participant recalled having once claimed to have slept with 15 million women.
- "I believed that aliens wanted to have my baby."

- (P) The participant once had a serious premonition of "a guy stabbing me." He said, "I saw him but he didn't stab me."
- (P) The participant had previously struggled with auditory hallucinations and paranoid delusions related to "people trying to kill me."

Deduction: paranoid delusion (out-group only)

Participant 17: (Missing identification sheet)

- The participant provided disjointed testimony. For example, when asked about a possible history of hallucinations, "I have memories of voices of kids at school."
- The chart reported that the participant had inappropriately claimed various ordinary documents as fraudulent. There was apparently a history of disorganized speech and talking to self — but no details.

Deduction: probable psychosis; unable to differentiate.

Participant 18: 52-year-old female

- According to the chart, the participant reported that dead animals could be resurrected if cigarettes, which are produced using animal byproducts, were disposed of near the bases of trees. She apparently heard voices on occasion.
- Chart notes said, "She can (telepathically) communicate to her ex-husband."
- Chart notes said, "Ms. X stated that God, the devil and Jesus smoke marijuana."
- Chart notes said that the participant had, "Reported seeing God's light."
- (P) The participant said that she had joined a group of "Jesus freaks" in the 1980s. Now, she feels that "God doesn't care for me."
- (P) The participant claimed to have had visions of "going to the moon and having a hotdog," and "coming back to the mud." She believes that this vision affects her future in some way.
- (P) The participant said, "I was always trying to get out of my body, to watch myself, but the medications stopped that."

Deduction: religious delusion

Participant 19: 49-year-old male

- The participant acknowledged that he had previously heard voices telling him what to do; however, he didn't elaborate.

- The participant said something that made no sense to me: "People would leave me in different directions."
- The participant acknowledged that he was frequently afraid that people (he did not know) would hurt him, including physically. Also, he believed that "people" (he did not know) from the North End in Winnipeg wanted him to financially support a baby.
- (P) "I believed in witchcraft and the occult, but no longer."
- (P) "I believe God has more power than Satan."
- (P) "I believed I got out of my body but I don't believe that any longer."
- (P) "I believe in UFO's, I had contact with a pole entity." I asked, "What is a pole entity?" He replied, "Diane X." He also said, "Take a walk on the boardwalk ... dims prims, dims prims ... it's an alien transmission." I asked, "Who is Diane X?" The participant responded, "She's the pole entity."
- (P) The participant said that once while he was in the woods, "I walked into a hatch [of a UFO] and ended up in my home, in my bed."
- (P) The participant also talked of "the churches of Asia and the nine guiding lights."

Deduction: religious delusion, paranoid delusion (out-group only)

Participant 20: 41-year-old female

- The participant alluded to periods of religious preoccupation, recalling having been very interested in *The Da Vinci Code*.
- The chart reported that the participant had a history of auditory hallucinations and disorganized thinking.
- (P) The participant said that she once felt she was possibly Mary or Jesus. Also, after seeing a certain man, she felt that Jesus had returned.

Deduction: religious delusion

Participant 21: 23-year-old male

- The participant said, "White people are cannibalizing natives" (but didn't know who these white people were).
- The participant said that he had heard a voice telling him he'll live forever.
- The participant perhaps acknowledged having seen visions.
- The participant said that "bible fanatics" believe that he had put a curse on them.

- The participant was worried that "white supremacists" want to "wipe out" Natives and their "religion."

Deduction: Probable religious delusion, paranoid delusion (out-group only)

Participant 22: 58-year-old male

- The participant felt that "people" were trying to kill him (but he didn't know these people).
- "I had delusions I was somebody else..." The participant was initially reluctant to tell me who he had thought he was, but then acknowledged he had felt he was Jesus.
- The chart reported a history of auditory hallucinations.
- (P) In the past, "I thought I had demons in me." The participant felt the devil could control him (however, he was sniffing glue during this time).

Deduction: religious delusion, paranoid delusion (out-group only)

Participant 23: 53-year-old female

- The participant once heard her father's voice yelling, "Don't hurt her."
- The participant recalled having experienced that both the radio and TV would communicate with her directly. She once recognized her daughter in a movie. In the scene, she was being followed and threatened.
- (P) The participant said that once during a "breakdown," "an angel touched my shoulder." I asked, "Was it real?" The participant replied that at the time, "to me it was real."
- (P) I asked about possible out-of-body experiences. The participant answered emphatically, "Yes! I left my body and ended up in a park ... later on, I found out it was the park my parents first met at." Also, another time, "My body left the bed and I watched my husband and myself in bed."

Deduction: religious delusion, paranoid delusion (out-group only)

Participant 24: 20-year-old male

The participant denied any history of delusions, hallucinations or religious preoccupation.

- The chart reported that the participant had told nursing staff that he had evil spirits inside him.

- The chart reported that the participant had inappropriately urinated in all sorts of bottles. On the ward, he had episodes of barking like a dog. He also had "worries about people."

Deduction: religious delusion, possible paranoid delusion (out-group only)

Participant 25: 39-year-old male

- The participant previously believed, "people were out to get me." "The government was watching me." He believed that, "police, paramedics would incarcerate me."
- The participant previously heard hallucinatory derogatory comments like, "You're no good." These voices were also "swearing at me."
- The participant has noticed many "coincidences" and has had premonitions.
- (P) When asked about possible unusual experiences, the participant said, "I would see a light as a kid."
- (P) The participant sometimes sees halos around objects.
- (P) The participant described what may have been Lilliputian hallucinations. He alluded to "the lights of little people."
- (P) The participant said that his grandmother, who had died, had come back as a "ghost." He claimed to have had a premonition of her death.
- (P) The participant saw a UFO on a Native reservation. He said, "there were lights following us, yelling at us, others would say it was a man but it wasn't."
- (P) The participant said, "I saw myself flying. I don't know if it was a dream. Sometimes I think I'm actually floating, it's weird, like I don't have any weight."

Deduction: magico-religious delusion, paranoid delusion (out-group only)

Participant 26: 24-year-old female

- The participant said that she sometimes hallucinates music (such as songs on the radio).
- The participant had been preoccupied with people "talking about me." She felt that these people could "strike" at any time. She said, "Maybe I've done something wrong?"
- The participant felt that people or an "evil" entity "controls me like a robot."

- The participant said she had previously seen a very menacing man in her yard. "My mother told me I was hallucinating." "His name was Viceroy." I asked how she knew his name and the participant replied, "Somehow, I just knew."

Deduction: magico-religious delusion, paranoid delusion (out-group only)

Going Back a Hundred Years

I couldn't help but wonder if this pattern of magico-religious and out-group paranoid delusions would be evident in patients from another epoch. I spent a day in the medical records department of the oldest psychiatric facility in the province of Manitoba.

With the arrival of the railroad in 1870s, Manitoba's population exploded — from 25,000 people in 1871 to about 250,000 in 1900. Two major psychiatric facilities were constructed in the late 1800s to accommodate the burgeoning population — the Manitoba Asylum (now the Selkirk Mental Health Centre) in the town of Selkirk, just outside Winnipeg, and the Brandon Asylum for the Insane (now the Brandon Mental Health Centre) in Brandon, serving the agricultural communities in the southwestern part of the province.

I examined fifty random charts from the late 1800s and early 1900s — forty from the Selkirk facility and ten from Brandon. (Documentation tended to be better in the Selkirk charts.) It was my impression that most patients fell into one of three categories: psychotic patients (bipolar disorder or schizophrenia), dementias, and severe depression characterized by either considerable psychomotor retardation or significant suicidal behaviors. (Solely thinking about suicide did not usually result in admission in those days.) Irrational violent behavior was sometimes a reason for admission.

Charts before World War I were almost always handwritten, in black ink, accompanied by very brief descriptions. Sometimes the only written entry describing a patient's behavior would be, "Inmate appears foolish." Other common descriptors were "lunatic," "crazy," and "insane."

Of the fifty charts, the ones I have summarized in the following section represent the twelve best descriptions of psychosis. The charts represent a period from 1897 to 1937. Recall that chlorpromazine, the first effective treatment for schizophrenia, was introduced in the mid-1950s. Marginal treatments like insulin coma treatments, electroconvulsive therapy and lobotomies were increasingly used during the 1940s. Other than providing a supportive milieu, there were no specific treatments noted in any of the charts.

Patient Delusions from Historical Records

Chart 1: 25-year-old male, admitted in 1897

This patient was diagnosed with dementia praecox, hebephrenic type, and tuberculosis. He was described as an "Indian." The chart said, "He has religious mania and he believes that the apparition of his deceased mother appears to him at night. His friends and neighbors are afraid of him."

Deduction: religious delusion

Chart 2: male, admitted in 1898

At admission, the patient was described as "acting peculiar," "walks in circles" and "his sister informs one that he will wash and dry his hands six times in succession."

It was also written that, "He entertains a great many delusions but does not react to them." Because this patient's behavior was not erratic, he was able to regularly work on a farm. However, while he was working, others suspected he regularly hallucinated. It was also written emphatically, "He is a confirmed masturbator."

Deduction: none

Chart 3: male, admitted 1902 (Brandon)

This patient was described as perpetually "nervous" and unable to work. The patient complained "the British government were after him to take him up in the air and drop him into molten babbitt [a metal alloy]."

Deduction: paranoid delusion (out-group)

Chart 4: 17-year-old female, admitted 1913

The chart noted that this patient had "acted strangely" and described a "condition of a maniac." It was also written, "Her delusions are varied. One yesterday was that they were trying to prison [sic] her."

Deduction: paranoid (probable out-group)

Chart 5: 34-year-old male, admitted 1917, ex-soldier

This patient was "suffering from delusional insanity with suicidal tendencies. Delusion consists in the belief that he has a petrified brain which is so hard that nothing can damage it." Otherwise, the patient talked rationally. During the war he was "being used to bring ammunition for a Stokes gun."

Prior to his admission, this man had put a pound of powder in a metal container and placed it to the back of his head. The ensuing explosion left him unconscious for several hours in the woods before a passerby found him. The explosion resulted in superficial scalp injuries and a right hand amputation. He was discharged eight months after admission.

Deduction: I would be inclined to classify a delusion of a "petrified brain" as a non-religious delusion (or, at most, weakly religious). A petrified brain does not appear to be a complete violation of humanness.

Chart 6: male, admitted 1919

The patient was described as an "Indian" who "laughs without cause. Says he goes walking with priests because no others have the courage to go."

Deduction: possible religious preoccupation

Chart 7: male, admitted 1919, ex-soldier

This patient was a Canadian soldier who appeared to have been wounded in 1919. The chart said, "on being wounded he passed through Field Ambulance CCS & various hospitals, no signs of mental trouble was observed until he went to No. 5 Canadian G.H. where he developed melancholia and moroseness. Refused food. Wanted to poison himself. Given to masturbation." "He has auditory hallucinations — states 'voices keep humming' in his ears — that they tell him to do things as take his life."

Deduction: Possible paranoid delusions, however, may be better classified as a nihilistic delusion

Chart 8: 40-year-old, male, admitted 1919, ex-soldier

"Has hallucinations of feeling has electricity applied to his head." The patient apparently also had auditory hallucinations that would tell him that he "died in France, and came to life again." He had the grandiose delusion that "he is wanted in England and France now to reconstruct affairs." The physician noted, "As a rule, not accessible to conversation..."

Deduction: Magico-religious delusion

Chart 9: age "under 30," male, admitted 1918, ex-soldier

A physician's medical history in 1921 noted that the patient "talks nonsense ... He imagines that he has some superior power, travelling millions of miles per minute. He sewed peculiar buttons on his coat, and imagines that this will have some effect on the police." He also "quickly answers 'NO' or 'I don't know' to everything." There was a piece of potentially relevant social history: the patient's mother died when patient was five years old. He was apparently still doing poorly three years after admission.

Deduction: magico-religious delusion

Chart 10: 30-year-old, male, farmer, admitted 1928

"Patient feels electricity all over his body. Sees people all the time, on the floor, ceiling, and wherever he happens to look ... Some people tell him to run fast, some hit him, and they seem to be continually telling him to do various and odd things." The patient was described as being "very restless and violent." He would sometimes tear off all his clothes. At other times he would be mute and not move for hours at a time.

Deduction: paranoid (probable out-group)

Chart 11: 35-year-old, male, 1931 medical note

This patient was diagnosed with "delusional insanity." His 1931 admission note said: "refuses to sleep or eat," "hears voices at night," "patient is inaccessible and seems indifferent to his environment," and "jumble of meaningless words or phrases." He also "believes he is followed by certain persons" — that he appeared not to know personally. It was also noted that the patient would sometimes laugh inappropriately.

Deduction: paranoid (out-group)

Chart 12: male, admitted 1937

The patient was described as an "Indian" from an isolated northern Manitoba reservation (Norway House). The physician's note said that the patient "...claimed he heard the church bells ringing, asked all to go to church with him. Talked in Cree which was not coherent nor sensible." Prior to admission, he "kicked down the door and ran out into the bush, was found some hours later by wife lying on grass and singing hymns. Knocked all stovepipes down, throws dishes at anyone near." He was also talking to himself and believed to be hallucinating.

Deduction: religious preoccupation

Summary of Historical Cases

By my account, four cases appeared to include out-group paranoid delusions, three cases had magico-religious delusions and one case showed religious preoccupation. Therefore, even in the face of extremely sparse documentation there is still some evidence that eight of twelve psychotic patients could have had spiritual or out-group paranoid delusions.

At the turn of the century, electricity was a new and seemingly magical force; and accordingly two patients expressed delusions about this technology. In the ensuing years, newer quasi-magical technologies such as airplanes (UFO's), radios, televisions, microwaves, and implanted devices would become the content of many delusions.

It is also notable that four cases of psychosis involved World War I soldiers, which implies that the psychological toll of that war was probably significant. Notice that two of the four soldiers lacked the usual bizarre delusions (5, 7) seen in schizophrenia, suggesting that they may have been suffering with psychotic depressions (and PTSD).

Conclusion

When one audits delusions and hallucinations, it is acknowledged that no obvious system of classification stands out. The recognition of certain common themes may suggest possible ways to organize delusions into categories such as persecutory, self-referential, having delusions of influence, nihilistic, grandiose, and religious. These categories are not mutually exclusive and their relative evolutionary importance is currently unknown. Our small pilot study suggests that two prospective delusional categories, out-group paranoid delusions and magico-religious delusions, may be more prominent than previously recognized. Moreover, the potential evolutionary benefit of these two delusional categories is supported by a theoretical evolutionary model: the shamanistic theory of schizophrenia.

– 11 –
Finishing Touches

"If things seem under control, you're not going fast enough."
— Mario Andretti

This story began with a simple question: How does schizophrenia, a condition with an appreciable heritable component, persist if those afflicted consistently have fewer progeny? There is no doubt that this problem, known as the schizophrenia paradox, is important. In fact, even if the suppositions embedded within the inquiry are mistaken and it turns out we are dealing with a trivial question, the schizophrenia paradox has been valuable because it has expanded the terms of reference in our mission to understand schizophrenia. It has led many researchers, like me, to explore vast tracts of uncharted territory concerning the origins of schizophrenia.

The renowned evolutionary geneticist Theodosius Dobzhansky once said, "Nothing in biology makes sense, except in the light of evolution" (as cited in Mayr 2001, 39). There is no reason to think that schizophrenia should be any different. Rather than think about schizophrenia within the confines of a few generations, the evolutionary perspective opens the door to thousands of years of potentially fertile exploration.

Darwin was prescient about the importance of evolutionary theory to mental conditions. In the final pages of the *On the Origin of Species*, Darwin wrote: "In the distant future I see open fields for far more important researches. Psychology will be based on a new foundation, that of the necessary acquirement of each mental power and capacity by gradation. Light will be thrown on the origin of man and his history" (1859). One of the most important consequences of evolutionary thinking is that it brings us back to hunting and gathering societies — the natural environment of our species. For hundred of thousands of years, our genes adapted to traditional life. Natural activities are often better for us than their modern version; walking is safer than skiing, pure water is healthier than soft drinks. Some facets of humankind's original setting can certainly be improved upon — electric heat is a marked improvement over noxious campfire smoke — but more often than not, traditional life embodies the parameters that are most compatible with our natural propensities. We cannot turn back time, nor should we endeavor to affix ourselves to the past, but learning about our earliest history can be informative and constructive.

Twenty-first-century life can be at odds with the natural cadence of our species. This is doubly true for those who have schizophrenia. There are many sad stories concerning schizophrenia in the context of Westernization. I do not even have to dig into my clinical caseload for examples; just this week, as I write these words, the Internet news describes a television actor who killed his mother while screaming biblical passages (Rosenberg 2010). One of my medical school classmates, who flunked out due to schizophrenia, took his life a few years later. My uncle, who was never ashamed that he had schizophrenia, spent his entire life unemployed despite having a college diploma. I would guess that almost every reader has similar knowledge of someone struggling with the pain of mental illness.

Yet schizophrenia does not necessarily have to be bleak. A number of promising remedies exist, such as medications, counseling, and modifying the surrounding social milieu. For example, here in my hometown of Winnipeg we have an art studio for patients with mental illness. It is the brainchild of Nigel Bart, an accomplished artist who has also been diagnosed with schizophrenia. A few years ago he gave me a tour of his 3500 square foot facility that on any given day has about a dozen artists working on creative projects. Nigel told me, "Artbeat began as a simple and somewhat naive idea of mine. I wanted to create a place for creative individuals living with mental illness, as I recognized that there was a strong connection between mental illness and creativity, primarily in my own healing process, and a gap in this type of service." Although it was not Nigel's intention to mimic the original evolutionary setting of schizophrenia, I believe that he has intuitively created a natural and comfortable environment for people with psychotic conditions. The success of Artbeat Studio (http://artbeatstudio.ca/) cannot be easily quantified, but I have had many patients who have participated and their time in the studio has always turned out to be a positive period in their lives. They have felt competent and useful, and found kinship within a community that understands them. Small battles are indeed won amidst this troublesome war.

I wish to conclude this book in two parts. The first part examines three straggling topics related to the shamanistic theory of schizophrenia: coordinating the theory with recent research on religious cognition, exploring the possible relationship between humor and religion, and integrating bipolar disorder into the shamanistic theory. The second section summarizes the major points of the theory.

Convergence with Research on Religion

Several researchers investigating the psychology of religion have begun to pay attention to the shamanistic theory of schizophrenia because of its compatibility to their lines of study (Rogers and Paloutzian 2006; Taber and Hurley 2007). For example, Taber and Hurley have made some intriguing insights into possible neuroanatomical connections between psychosis and religion. They begin with the premise that delusions are a "misattribution of one's own actions or thoughts to an external agent" and that a similar cognitive misattribution is associated with mythical agents. Taber and Hurley refer to a number of fMRI studies implicating the anterior insula as the area of the brain that attributes action to the self, whereas the inferior parietal lobe may be involved in attributing actions to other agents. Drawing evidence from several neurocognitive studies, Taber and Hurley hypothesize that the neuronal circuitry linking these two brain areas may be integrally involved in both schizophrenia and religion.

Another line of inquiry proposes a shared psychological mechanism between delusions and religious ideation (Brüne 2009). For a number of years, evolutionary psychiatrist Martin Brüne has been studying the inner workings of delusions. He has focused his studies on people with delusional disorders, which are disorders in which the patient has non-bizarre delusions not caused by drugs or other illnesses. Such patients, by definition, possess delusions untainted by additional schizophrenic symptoms. This enables the study of delusions in their "pure" state. Brüne outlines, in his typical methodical manner, how certain "cognitive biases" observed in schizophrenia also apply to religious thinking. Expanding on the pioneering work of Chris Frith (2005a, 2005b), Brüne outlines four types of cognitive pitfall that are characteristically observed in both delusional patients and staunchly religious people: deficits in certain types of reasoning, difficulties with judging causality, compromised theory of mind (the ability to imagine alternate perspectives) and a diminished sense of self as an agent (Brüne 2009, 222).[1]

In a recent paper, Dein and Littlewood (2011) have argued that a common evolutionary trajectory may have been shared by both religion and schizophrenia. After dissecting Pascal Boyer's cognitive structure of religion (see Chapter Nine) and working backwards, the authors propose that theory of mind and the ability to attribute intention to entities (i.e., animals and humans) could have been the prerequisite cognitive skills for the development of religious thinking. Religion and schizophrenia may somehow be related because both involve over-attributions of inten-

---

[1] Although Brüne is aware of the shamanistic theory (and it would fit nicely into his hypothesis), he sidesteps mention of it due to (I suspect) his reservations about the relevance of group selection to *Homo sapiens*.

tion (e.g., paranoid psychosis, or the existence of gods which cannot be verified).

Humor and Religion: A Clandestine Relationship

Not only do humor and religion seem unrelated, the lightness of humor is in some ways at odds with the seriousness of spirituality (Polimeni and Reiss 2006). A subterranean link could, however, exist between these cognitive systems. Let me explain by first outlining the cognitive structure of humor.

Thomas Veatch has authored perhaps the clearest general cognitive theory of humor (Veatch 1998). According to Veatch, humor contains two elements — incongruous to each other — of which one is in "violation" of the "subjective moral order." (That is why sex and fart jokes are so funny, and why highly pious people have difficulty making people laugh.) Veatch's equation seems to work for every joke encountered. For example:

> Mommy, Mommy! What is a delinquent child?
> Shut up and pass me the crowbar.

The first element (set up by the child's innocent question) is that the wrongness of stealing is socially arbitrary and not necessarily obvious to a social newcomer. The second element is mother's incongruous reply, which contains the social violation — mothers are supposed to disapprove of rather than encourage criminal behavior.[2]

Examine any joke according to Veatch's scheme and you'll find similar incongruencies — one of which will always hint at a social violation. When the mind recognizes such a scheme, a brief involuntary seizure called laughter is sometimes elicited.

The fascinating thing is that the same general pattern seems to be shared by religious ideation. Recall that a religious idea — according to Pascal Boyer — is the product of a violation of an ontological category. In other words, formulate a cognitive scheme, contradict it, and wait for an emotional reaction — transcendental feelings in one case, laughter in the other. This process (of violating or contradicting cognitive schemes) could be rooted in our ability to utilize syntactical negation: simply add "not" to a sentence and you turn the entire idea around. However, grammatical negation is a signal and does not per se stimulate an emotion. To my knowledge (and admittedly I may be missing other examples), only

---

[2]Jokes are indeed complex and this interpretation is clearly an oversimplification. A good joke can implicate several social attitudes and rely on a myriad of subtle word associations — a misplaced word will often kill a joke. Moreover, there may be several indistinct layers of incongruous and resolving elements in any given humorous stimuli.

religion and humor are candidates for this psychological mechanism. This idea is speculative, but it should be kept in back of mind. Evolution is an old pro at recycling existing structures for entirely new purposes — the tongue, for example, is used for speaking as well as eating. What would have come first — humor or religion? Based on anecdotes of teasing between primates, I would place my bets on humor.

How Does Bipolar Disorder Fit In?

Bipolar disorder (also known as manic depression) is a psychiatric condition characterized by manic episodes alternating with depressive periods. Typical symptoms of mania include euphoria, increased energy, rapid speech, flight of ideas, insomnia, irritability, capriciousness and hypersexuality. Manic episodes can last several weeks or months, and patients are frequently, but not necessarily, psychotic. The proportion of manic episodes compared to depressive moods varies in each individual. Although a sizeable number of people with bipolar disorder exhibit persistent idiosyncratic thinking (which prevents integration into the workforce), they can often live symptom-free for years.

The parameters underlying the schizophrenia paradox are equally applicable to bipolar disorder: the existence of bipolar disorder is an evolutionary enigma in precisely the same way as schizophrenia (Wilson 1998). Moreover, bipolar disorder is symptomatically similar to schizophrenia. This prompts the question: how is bipolar disorder evolutionarily related to schizophrenia?

The most compelling similarity between bipolar disorder and schizophrenia is that both conditions involve (non-delirial) psychosis — a phenomenon that is otherwise rare in nature. In addition, many symptoms partially overlap. Recall from Chapter Four that a number of neurocognitive findings associated with schizophrenia are sometimes evident (albeit less robustly) in bipolar disorder (Ketter et al. 2004; Martin et al. 2007). From a clinical perspective, add a little moodiness and a non-bizarre delusion to a schizophrenia case and you begin to have real problems distinguishing these diagnoses.

Although many similarities exist, there are a few key differences. Foremost, each diagnosis runs predominantly along different family lines (Cardno and Murray 2003; Lapierre 1994; Loranger 1981; Smoller and Finn 2003; Somnath et al. 2002). These lineages are not completely separate, as some families show both diagnoses (Ketter et al. 2004). Another distinction involves lithium treatment. Although a number of medications are effective for both schizophrenia and bipolar disorder, lithium frequently blunts mania but is hardly ever effective in classic schizophrenia.

So what about spiritual delusions? Are magico-religious delusions also found in bipolar disorder? In the only study I could find that compared conventional religious delusions between the two diagnoses (Applebaum et al. 1999), the results were surprisingly similar: 35 percent in schizophrenia and 33 percent in bipolar disorder. (They also tallied religious delusions in psychotic depression (14 percent) and substance-abuse psychosis (17 percent).) Some minor discrepancies between schizophrenia and bipolar disorder were demonstrated in three other delusional categories that we know may reflect magico-religious delusions. The results are shown in Table 11.1.

	Body/mind control delusions	Thought broadcasting	Persecutory delusions
Schizophrenia	75%	52%	83%
Bipolar Disorder	49%	18%	71%
Depression	36%	16%	73%
Alcohol/Drug Abuse	50%	23%	73%

Table 11.1: Comparison of religious delusions in schizophrenia and bipolar disorder, as well as in cases of depression and substance abuse.

From an evolutionary point of view, why should bipolar disorder exist at all? My first idea was that bipolar disorder was some sort of "new and improved" offshoot of schizophrenia, but this didn't sit well with me. Wouldn't selection pressures simply have made a wholesale transformation from schizophrenia to bipolar disorder? Perhaps, we, in the twenty-first century, were coincidentally catching evolution halfway through this transformation process — possible, but not likely. I also considered whether the evolutionary advantages of mania had something to do with enhancing humankind's creativity and that any similarities to schizophrenia were just coincidence. This too seemed unlikely — why would bipolar disorder have almost as many spiritual delusions as schizophrenia? Besides, there would probably be simpler ways for evolution to crank up hominid creativity.

Then, one evening, a possible solution dawned on me. I happened to be at a party where I knew most of the people and I began to notice that exuberant people were coupled with equally effervescent spouses, while reserved people tended to be with similarly shy partners. I began to think, what if shamanistic personalities intermarried along two lineages — people full of affect versus relatively reserved people? Over thousands of years, could such assortative mating create two varieties of shamanistic personality? Larger civilizations, with their greater opportunity to choose mates similar to oneself, would only exacerbate the situation. And recall, Kraepelin's primary division between schizophrenia and bipolar disorder was affect-ridden psychosis versus non-affect-ridden psychosis. Let's examine this idea more closely.

Human mating is not random — there is evidence that, for certain personality traits, like attracts like. Although research into marital compatibility is in its infancy, a few general themes are beginning to emerge. Assortative mating appears to occur in the human species for such broad traits as physical stature and intelligence (Epstein and Guttman 1984; Reynolds et al. 1996; Silventoinen et al. 2003). Social attitudes such as conformity (Epstein and Guttman 1984), religiosity and political opinion (Epstein and Guttman 1984; Luo and Klohnmen 2005) also tend to be comparable in spouses while personality traits such as extroversion and neuroticism have yielded equivocal results (Epstein and Guttman 1984; Hur 2003; Luo and Klohnen 2005; Mascie-Taylor 1989). Interestingly, personality traits associated with bipolar spectrum conditions such as sensation-seeking, disinhibition and boredom susceptibility seem to aggregate within certain couples (Glicksohn and Golan 2001).

In schizophrenia, a number of studies have shown a tendency for assortative mating. In one study, over one-third of spouses of schizophrenic patients revealed schizophrenia-spectrum symptoms (Rosenthal 1975). Another study with a low sample size (thirty couples) showed 6 percent of schizophrenic spouses also having schizophrenia (Alanen and Kinnunen 1975). Parnas's (1988) analysis of the Copenhagen High Risk data set showed 13 percent of spouses of schizophrenia patients having Cluster A personality traits (schizophrenia-like traits) compared to 2 percent for the control group. Although these studies generally had low sample sizes, collectively they demonstrate distinct tendencies for assortative interactions in schizophrenia.

Assortative interactions have also been shown to occur in bipolar disorder (Mathews and Reus 2001; Merikangas and Spiker 1982). Furthermore, such assortative mating may be more pronounced in bipolar disorder (and schizophrenia) compared to other psychiatric conditions. For example, two studies have shown that assortative interactions are more pronounced in bipolar disorder than in unipolar depression — a less heritable condition (Mathews and Reus 2001; Merikangas and Spiker 1982).

As anyone who has ever worked on a psychiatric inpatient ward can attest, patients often socialize along certain diagnostic clusters. The bipolar-schizophrenia division may simply be the result of affect-spectrum assortative mating being superimposed on psychotic-spectrum assortative interactions.

A Final Synthesis of the Shamanistic Theory of Schizophrenia

Everyone, at some point, has wondered about the maelstrom of insanity — a set of baffling experiences and indecipherable behaviors associated with a small segment of humanity. The loose confederation of mystifying actions we now call schizophrenia has been present through many epochs and every locality. The schizophrenia syndrome is by no means uniform — substantial variability can be found in its "symptom" presentation. In addition, there is always a slice of the population that isn't afflicted in full, but experiences a lesser version of this phenomenon. But the bottom line is that, for some time now, something out of the ordinary has been occurring in a small proportion of young adults.

Other common psychiatric illnesses have some sort of rational explanation within reach. Depression appears to have something to do with mammalian expressions of submission, while anxiety is connected to fear and fleeing. Obsessive-compulsiveness may be linked to overly persistent habits, and mental retardation is the result of incomplete brain development. In contrast, schizophrenia is not an obvious extension of any other behavior and has no tangible analogue in nature. The motor and sensory systems have gone awry but not in a haphazard or unsystematic way. Clear patterns exist but are inexplicable in the context of twenty-first century life. Despite two centuries of scrutiny, mainstream psychiatry has yet to tell us anything particularly insightful about the condition. Schizophrenia seems too elusive for twenty-first century science, perhaps because the psychiatric establishment keeps attacking the problem from the same angle, trying to put a square peg in a round hole.

The best available evidence suggests that schizophrenia is a heritable condition that has been around for many generations, which means that incorporating evolutionary theory into the discourse is not an option, but a requirement. Although basic evolutionary theory does not unequivocally disprove disease models of schizophrenia, it does imply that something more may be happening. Consistent with this is the observation that schizophrenia does not fit neatly under the rubric of classic neuropsychiatric diseases. As we examine such topics as anthropology, population dynamics and genetics, an alternate explanation of schizophrenia begins to emerge.

The resemblance of schizophrenia to the ancient institution of shamanism surpasses coincidence. The personal histories of shamans so often contain episodes of serious mental disturbance, including numerous accounts of spiritual hallucinations. Shamans become engrossed with magico-religious beliefs and suspicions of otherworldly threats. They have the reputation of being solitary, unsociable and odd. Just like schizophrenia, which exists in every part of the globe, shamanism has been practiced on every continent. Every known traditional society is known to have carved out a dedicated role for such magico-religious practitioners, and up until 10,000 years ago, all humans lived in traditional societies. It is therefore likely that for tens of thousands of years, shamans were present wherever humans roamed the earth. Where schizophrenia is common, magico-religious healers are not — and where shamans practice, insanity becomes spiritual.

There is no way to be absolutely certain that the rates of shamanism line up perfectly with schizophrenia's 1 percent prevalence; however, all the best evidence suggests that it must be close. Rates of shamanism seem to vary between about 0.5 percent to — the occasional tribe claims — about 10 percent. (It depends on how much one extends the definition of shamanism. Recall that hallucinogens can make almost anybody into a temporary shaman.) Most tribes contain, give or take, about a hundred people and tribes usually have one practicing shaman. Shamans frequently, but not always, have histories of psychosis (and sometimes a psychotic person does not become a shaman). All in all, through history, psychotic-prone shamans have probably comprised somewhere around 1 percent of the population (adding bipolar disorder would make this figure 1.5–2 percent).

The universality of shamanism is one of those fascinating anthropological tidbits that upon further reflection has monumental implications. How could every tribe across the globe independently create such a highly specific social role? It would be like every multinational company in the world posting the same job description:

> Position available for person who shall heal the sick, divine in arbitrary matters, lead magico-religious rituals and prognosticate about war. Applicants having had a psychotic episode in young adulthood will be given special consideration.

One would conclude that they copied from each other — but we know that cultural diffusion cannot explain the universality of shamanism. Besides, it is rare for arbitrary cultural traditions to maintain near-perfect fidelity for more than a few hundred years. (Just look at how much languages change in a few hundred years.)

In hunting and gathering societies, the universality of such roles as hunter, warrior or chief is partially based on their objective value. When faced with hunger, it is sensible to pick up the sharpest implement at your disposal and kill an animal. Similarly, it is obvious that showing deference towards a patriarchal figure can help achieve complex tasks or resolve disputes. In contrast, shamanism is not a direct solution to the evolutionary dilemma it seems to solve. It is hard to figure that traditional people, even on an intuitive level, would have realized that shamans move tribes forward. Shamanism makes evolutionary sense but only in a very circuitous way, seemingly dictated by certain constraints of brain evolution — the tribal advantages are simply too abstract to be appreciated by the participants. For all these reasons, shamanism would have to be genetically primed. And indeed, shamanism — like mental illness — usually overcomes select individuals.

Religion can be considered a form of specialized communication, similar to physical gestures, language, music or humor. In the animal kingdom, communication is often asymmetrical, meaning that some members of the species are better equipped to produce the signal while others receive it (e.g., male songbirds, the waggle dance of forager bees). In humans, there may be similar differentiation for certain communication skills like humor and music. In the case of religion, shamans are adept at communicating magico-religious ideas while the rest of humanity, to varying degrees, can receive their messages. The resulting communication changes the recipients' behavior: perhaps escalating vigilance towards other tribes, or increasing participation in group-building ceremonies.

Shamans heal, divine, lead rituals, advise about matters of war and create religious content. Some of these assets may be more evolutionarily important than others, and some may even be immaterial. Shamans mostly concern themselves with religion, which facilitates social cohesion through a variety of loosely related mechanisms. For example, spiritual rituals are part of a system that allows arbitrary decisions — on matters such as where to hunt, when to plant crops or whether to attack a neighboring tribe — to be unanimously supported by the community. Religion can also help affirm social roles, which allows tribes to function more effectively through division of labor. Because the transition between life phases — childhood, adulthood, spouse, elder — can be indistinct, rituals serve to demarcate social positions and back up the tribe's expectations for each role with the weight of spiritual ceremony.

Still another shamanistic asset could be the tendency to fixate on otherworldly dangers. Paranoid delusions generally center on agents outside one's community, and therefore become linked to intertribal warfare. In the field, shamanistic prophecies can increase tribal vigilance or promote surprise attacks — highly effective assets for group competition.

In traditional societies, shamanistic ramblings will eventually develop into religious ideologies. If the shamanistic theory is correct, we should see some resemblance between psychosis and religious doctrine; and in fact, the general form of psychotic delusions seems to conform to the deep cognitive structure of religious ideation.

The study of new religious movements is a sort of artificial laboratory serving as a window into the ancient world of tribal shamanism. Cults are probably a cultural quirk resulting from humankind's propensity towards shamanism manifesting within our large stratified civilization. Cults are formed when three stars align: 1 — a charismatic personality endowed with 2 — mildly delusion ideas represents 3 — a marginalized segment of society. Parenthetically, a good proportion of this marginalized segment may be excluded from mainstream society precisely because they exhibit small measures of psychosis (especially idiosyncratic or disordered thinking). In large, complex societies, this triad is the fountainhead of new religions. Such a potent combination could have occasionally been present in traditional societies; however, the typical extremism so often seen in cults would have normally been tempered by skeptics in the tribe. Accordingly, anthropological accounts show that although shamans were customarily influential in group-level decisions, they were by no means the final word.

One of the more radical implications of the shamanistic theory is that shamans are their own specific genotype apart from the rest of society. In other words, human groups may be analogous to honeybee colonies comprised of task specialists: forager bees, guard bees, and so on, each genetically programmed with a different set of behavioral instructions. Although this idea may sound like science fiction, it is not entirely radical. A similar situation can be observed in social hierarchies, which are common in predatory mammals that hang out in groups (e.g., hyenas, wolves, primates, hominids). The different hierarchal behaviors are partially determined by genes and biological constitution (e.g., serotonin levels). I would argue that status hierarchies are evolutionarily advantageous — at a group level — because they form a sort of "soft" division of labor. Although the hierarchical transitions are gradual, the net behavioral effects result in slightly different social roles for each animal.

Although the gene-centered view of evolution has merit, it has skewed our perspective of how genes are distributed throughout a population. Moreover, most genetic equations and evolutionary models presume — for the sake of simplicity — random interactions between organisms. However, the idea that genes are independent of each other does not seem to be the case, especially among animals that form tight-knit groups. Imagine a grid with a hundred boxes, each square representing an organism in a population. Nature does not splash genes upon the grid randomly. Instead assortative mating, hybrid vigor and heterozygote advantage skew the splotches, creating aggregations of genes which can theoretically create advantages or disadvantages for the entire grid — that is, the group. In this jostle of evolution, certain gene combinations will achieve a modicum of stability, especially if they help a part of the grid — groups — survive. The complexity of the environment and all its selection pressures creates a diversity of pushes and pulls on any given gene, which makes it difficult to predict how this all "should" turn out in the end. But though it may be tough to retrace all the steps, we are privy to the final result: evolutionarily programmed division of labor frequently occurs in animals that compete as groups.

However, task specialization in hominids and other mammals is not as well circumscribed or categorical as in, say, the eusocial insects. This is because evolutionary progression in hominoids has mostly centered around social behaviors, which tend to be subtle and nuanced. In turn, social behavior appears to be shaped by polygenic mechanisms, thereby creating a variety of behaviors. Although shamanism and schizophrenia are both identifiable entities, the phenomenon can create a spectrum of intermediate forms which blend into normality (the dimensional model of schizophrenia) (Peralta and Cuesta 2005).

Those who argue for a categorical approach to schizophrenia contend that there is a point of rarity on the bell shaped curve — meaning that schizophrenia is not the last 1 percent of the human behavior bell curve, but a smaller blip protruding beyond the end of the normal tail. In my view, it is premature to dismiss categorical approaches to schizophrenia. Although some schizophrenia symptoms (e.g., autism, affective flattening, avolition) may show dimensional properties within a population, other characteristics seem more distinctive. It is possible that the ultimate behavioral effect of psychosis is categorical in practice. For example, people either get fixated on delusions (and act on them) or they sweep them under the rug. But recall the patient who wasn't sure he was Jesus; perhaps *he* is that rare intermediary form.

Task specialization is compatible with polygenic modes of inheritance (the probable genetic scenario in schizophrenia). There may be a certain threshold number or combination of schizophrenia genes that triggers certain physiological processes. Suppressing the schizophrenia phenotype below a certain number of genes or activating it in other genetic or environmental circumstances could give schizophrenia that categorical feel. Such epigenetic mechanisms theoretically exist and could very well apply to schizophrenia (remember the stress diathesis model of schizophrenia).

We also know that there are at least two evolutionary mechanisms (heterozygote advantage and assortative mating) that could lead to task specialization, and both mechanisms could be applicable to schizophrenia. In keeping with the heterozygote advantage model, people with a touch of schizophrenia seem to enjoy special assets. Recall that schizotypal people seem to show enhanced creativity, and relatives of schizophrenic patients demonstrate increased intelligence. Assortative mating, which is also seen in schizophrenia, can theoretically work in tandem with heterozygote advantage to support task specialization.

Is schizophrenia a disease? I believe the answer is both yes and no. From an evolutionary perspective, schizophrenia may not be a disease, but it is burdensome to the individual. Schizophrenia may very well be like sickle cell anemia, that other well-known heterozygote-advantage condition. In a malaria-infested environment, a sickle cell heterozygote is advantageous;

however, in the absence of malaria, you are better off having nothing to do with sickle cell genes. Similarly, in the absence of hunting and gathering subsistence, schizophrenia is evolutionarily meaningless. On a personal level, modern life usually makes schizophrenia a setback, sometimes a tragedy and rarely, if ever, beneficial.

The many "nature versus nurture" inputs underlying human behavior are still being worked out — not to mention how (or whether) genes interact with cultural institutions (Boyd and Richerson 2002). The establishment of a learned behavior (or cultural institution) can hypothetically have effects on genes. The Baldwin effect is a theoretical situation whereby a learned trait can eventually become instinctual. Let's say our ancient hominid ancestors learned one day that throwing a spear is a good way to procure food. Those hominids with better "throwing genes" may have supplanted poor throwers, possibly encouraging the evolution of genes that eventually make the behavior instinctual. Although we may recognize some baseball pitchers as being "naturals," all humans are natural throwers compared to other animals. Throwing may be, to some extent, instinctive to our species because we are all genetic descendants of prehistoric spear-throwing hall of famers. A Baldwinian mechanism might contribute towards the evolution of cultural institutions like shamanism. For example, if the community "learned" to appreciate some aspect of shamanism, it may have preferentially supported those associated behaviors, which in turn would have accelerated the evolution of shamanism.

If shamans and schizophrenia seem superficially different, it is because they occupy completely different socio-cultural spaces. Just as we may not be able to imagine a person with schizophrenia being an influential and respected community member, many eighteenth-century slave owners could not have imagined a black person capable of being the president of the United States of America. In their own tribal communities, shamans were often considered odd and "crazy," but because they could also be productive citizens, they were seen in a completely different light.

The Western world has a tendency to gloss over the negative consequences of its technologies and impersonal bureaucratic institutions. The rigid social expectations which industrialization introduced, whether in strait-laced offices or on clockwork-driven factory floors, are incompatible with the shamanistic mind; but shamanistic personalities are the proverbial canary in the coal mine. Even "normal" people have trouble adjusting to the demands of industrialized society. In the nineteenth century, a number of intellectuals such as Franz Kafka, Karl Marx, Émile Durkheim and Max Weber began to recognize these problems and wrote thoughtfully about the inflexibility and depersonalization of modern bureaucracies. It is no coincidence that this is the same period that saw the first "epidemic" of schizophrenia.

The secularization of Western society has further curtailed magico-religious thinking and made shamanistic personalities obsolete. A few centuries ago, superstitious beliefs such as divine miracles, demonic curses or magical concoctions could be discussed within the context of daily events. In the twenty-first century, this kind of spiritual discourse has become the domain of cults, paranormal societies and churches. Even established religions have become, in a way, secularized. Although every major religion began with magico-religious superstitions, these institutions now tend to downplay such stories and increasingly treat such ideas as metaphors.

Like the perfect storm, everything went wrong for the shamanistic personality over the last few centuries. Industrial changes undeniably have a huge plus side, but they can also marginalize certain segments of society. As magico-religious thinking grew increasingly incompatible with modern society, the shamanistic personality became the canary in the coal mine. We can, however, retrace our steps and see how the downtrodden of one era may have prospered in a different setting. It may take a bit of imagination, but I find it possible to envision most people with schizophrenia as shamans. If such an evolutionary course did take place, it is of utmost importance to explore the route — for every additional fact about schizophrenia brings us one step closer to controlling it.

References

Abraham, A., S. Windmann, R. Siefen, I. Daum and O. Gunturkun (2006). "Creative thinking in adolescents with attention deficit hyperactivity disorder (ADHD)." *Child Neuropsychology* **2**: 111-123.

Ackerknecht, E. (1943). "Psychopathology, Primitive Medicine, and Primitive Culture." *Bulletin of the History of Medicine* **14**: 30-67.

Ackner, B., A. Harris and O. A.J. (1957). "Insulin Treatment of Schizophrenia: a controlled study." *Lancet* **272**(6969): 607-611.

Adachi, N., M. Matsuura, T. Hara, Y. Oana, Y. Okubo, M. Kato and T. Onuma (2002). "Psychoses and epilepsy: Are interictal and postictal psychoses distinct clinical entities?" *Epilepsia* **43**(12): 1574-1582.

Alanen, Y. O. and P. Kinnunen (1975). "Marriage and the development of schizophrenia." *Psychiatry* **38**(4): 346-365.

Albertson, D. N. and L. E. Grubbs (2009). "Subjective effects of Salvia divinorum: LSD- or marijuana-like?" *Journal of Psychoactive Drugs* **41**(3): 213-217.

Aleman, A., R. S. Kahn and J. P. Selton (2003). "Sex Differences in the Risk of Schizophrenia: Evidence from Meta-analysis." *Archives of General Psychiatry* **60**: 565-571.

Alexander, F. G. and S. T. Selesnick (1966). *The History of Psychiatry: An Evaluation of Psychiatric Thought and Practice from Prehistoric Times to the Present*. New York, Harper and Row.

Alexander, R. (1987). *The Biology of Moral Systems*. New York, Aldine de Gruyter.

Allen, C. (1975). "The Schizophrenia of Joan of Arc " *History of Medicine* **6**(3-4): 4-9.

Allison, A. C. (1954). "Protection afforded by sickle-cell trait against subtertian malarial infection." *British Medical Journal* **1**(4857): 290-294.

Aloia, M. S., M. L. Gourovitch, D. Missar, D. Pickar, D. R. Weinberger and T. E. Goldberg (1998). "Cognitive Substrates of Thought Disorder, II: Specifying a Candidate Cognitive Mechanism." *American Journal of Psychiatry* **155**(12): 1677-1684.

American Psychiatric Association. (1994). Diagnostic and statistical manual of mental disorders (4th ed.). Washington, DC: Author.

American Psychiatric Association. (1980). Diagnostic and statistical manual of mental disorders (3rd ed.). Washington, DC: Author.

Anderson, K. E., T. A. Linksvayer and C. R. Smith (2008). "The causes and consequences of genetic caste determination in ants (Hymenoptera: Formicidae)." *Myrmecological News* **11**: 119-132.

Andreasen, N. C. (2000). "Schizophrenia: the fundamental questions." *Brain Research Reviews* **31**: 106-122.

Andreasen, N. J. and P. S. Powers (1975). "Creativity and Psychosis." *Archives of General Psychiatry* **32**(1): 70-73.

Andreone, N., M. Tansella, R. Cerini, G. Rambaldelli, A. Versace, G. Marrella, C. Perlini, N. Dusi, L. Pelizza, M. Balastrieri, C. Barbui, M. Nose, A. Gaspirini and P. Brambilla (2007). "Cerebral Atrophy and White Matter Disruption in Chronic Schizophrenia." *European Archives of Psychiatry and Clinical Neuroscience* **257**(1): 3-11.

Angst, J. (2002). "Historical Aspects of the Dichotomy Between Manic-Depressive Disorders and Schizophrenia." *Schizophrenia Research* **57**(1): 5-13.

Applebaum, P. S., P. C. Robbins and L. H. Roth (1999). "Dimensional Approach to Delusions: Comparison Across Types and Diagnoses." *American Journal of Psychiatry* **156**: 1938-1943.

Ariely, D. (2008). *Predictably Irrational: The Hidden Forces That Shape our Decisions.* New York, HarperCollins.

Arnold, C. (2008). *Bedlam: London and its Mad.* London, Pocket Books.

Ashley-Koch, A., Q. Yang and R. S. Olney (2000). "Sickle hemoglobin (Hb S) allele and sickle cell disease: A HuGE Review." *American Journal of Epidemiology* **151**(9): 839-845.

Atran, S. (2002). *In Gods We Trust: The Evolutionary Landscape of Religion.* Oxford, Oxford University Press.

Azari, N. P., J. Missimer and R. J. Seitz (2005). "Religious Experience and Emotion: Evidence for distinctive cognitive neural patterns." *The International Journal for the Psychology of religion* **15**(4): 263-281.

Azari, N. P., J. Nickel, J. Wunderlich, M. Niedeggen, H. Hefter, L. Tellmann, H. Herzog, P. Stoerig, D. Birnbacher and R. J. Seitz (2001). "Neural correlates of religious experience." *European Journal of Neuroscience* **13**: 1649-1652.

Baethge, C., P. Salvatore and R. J. Baldessarini (2003). "'On Cyclic Insanity' by Karl Ludwig Kahlbaum, M.D.: A Translation and Commentary." *Harvard Review of Psychiatry* **11**(2).

Bainbridge, W. S. and R. Stark (1980). "Scientology: To Be Perfectly Clear." *Sociological Analysis* **41**(2): 128-136.

Baird, J. (2008). *The Private John Lennon: The Untold Story from His Sister.* Berkeley, CA, Ulysses Press.

Bandelow, C., J. Krause, D. Wedekind, A. Broocks, G. Hajak and E. Ruther (2005). "Early traumatic life events, parental attitudes, family history, and birth risk factors in patients with borderline personality disorder and healthy controls." *Psychiatric Research* **134**(2): 169-179.

Barash, D. P. and J. E. Lipton (2001). *The myth of monogamy: fidelity and infidelity in animals and people.* New York, W.H. Freeman.

Bark, N. (2002). "Did schizophrenia change the course of English history? The mental illness of Henry VI." *Medical Hypotheses* **59**(4): 416-421.

Barkow, J. H., L. Cosmides and J. Tooby (1992). *The Adapted Mind: Evolutionary psychology and the generation of culture*. New York, Oxford University Press.

Baron, M., R. Gruen, L. Asnis and J. Kane (1983). "Familial relatedness of schizophrenia and schizotypal states." *American Journal of Psychiatry* **140**: 1437-1442.

Baron-Cohen., S. (2006). "Two new theories of autism: hyper-systemising and assortative mating." *Archives of Disease in Childhood* **91**: 2-5.

Baron-Cohen, S., P. Bolton, S. Wheelwright, V. Scahill, L. Short, G. Mead and A. Smith (1998). "Autism occurs more often in families of physicists, engineers and mathematicians." *Autism* **2**: 296-301.

Barrantes-Vidal, N. (2004). "Creativity and Madness Revisited from Current Psychological Perspectives." *Journal of Consciousness Studies* **11**(3-4): 58-78.

Bassiony, M. M., M. S. Steinberg, A. Warren, A. Rosenblatt, A. S. Baker and C. G. Lyketsos (2000). "Delusions and hallucinations in Alzheimer's disease: prevalence and clinical correlates." *International Journal of Geriatric Psychiatry* **15**(2): 99-107.

Baumann, G., M. A. Shaw and T. J. Merimee (1989). "Low levels of high-affinity growth hormone-binding protein in African pygmies." *New England Journal of Medicine* **320**(26): 1705-1709.

Bearzi, M. and C. B. Stanford (2008). *Beautiful Minds: The Parallel Lives of Great Apes and Dolphins*. Cambridge, MA, Harvard University Press.

Beattie, J. (1776). *Essays: on the nature and immutability of Truth, in apposition to Sophistry and Scepticism, On Poetry and Music, as they affect the Mind, On Laughter, and Ludicrous Composition, On the Utility of Classical learning*. Edinburgh.

Beauregard, M. and V. Paquette (2006). "Neural correlates of a mystical experience in Carmelite nuns." *Neuroscience Letters* **405**: 186-190.

Belsky, J. (1997). "Variation in Susceptibility to Environmental Influence: An Evolutionary Argument." *Psychological Inquiry* **8**(3): 182-186.

Belsky, J., M. J. Bakermans-Kranenburg and M. H. van Ijzendoorn (2007). "For Better and For Worse: Differential Susceptibility to Environmental Influences." *Current Directions in Psychological Science* **16**(6): 300-304.

Berger, P. (1999). *Forever Showtime: The checkered life of Pistol Pete Maravich*. Dallas, Texas, Taylor Publishing.

Berrios, G. E. (1997). "The Origins of Psychosurgery: Shaw, Burckhardt and Moniz" *History of Psychiatry* **8**(29 pt 1): 61-81.

Berrios, G. E. and R. Hauser (1988). "The Early Development of Kraeplin's Ideas on Classification: a conceptual history." *Psychological Medicine* **18**: 813-821.

Berrios, G. E. and K. Kahlbaum (2007). "The Clinco-Diagnostic Perspective in Psychopathology." *History of Psychiatry* **18**(2): 231-245.

Bertranpetit, J. and F. Calafell (1996). Genetic and geographical variability in cystic fibrosis: evolutionary considerations. *Variation in the human genome*. D. Chadwick and G. Cardew, Wiley: 97-118.

Bhatia, T., M. A. Franzos, J. A. Wood, V. L. Nimgaonkar and S. N. Deshpande (2004). "Gender and procreation among patients with schizophrenia." *Schizophrenia Research* **68**: 387-394.

Bhugra, D. (1992). "Psychiatry in ancient Indian texts: A review." *History of Psychiatry* **3**: 167-186.

Bhugra, D. and S. G. Potts (1989). "Remission of Psychotic symptoms after Burn Injury." *British Journal of Psychiatry* **155**: 717-718.

Bickler, S. W. and A. DeMaio (2008). "Western Diseases: current concepts and implications for pediatric surgery research and practice." *Pediatric Surgery International* **24**: 251-255.

Birchler, J. A., D. L. Auger and N. C. Riddle (2003). "In Search of the Molecular Basis of Heterosis." *The Plant Cell* **15**: 2236-2239.

Birket-Smith, K. (1957/1960). *Primitive Man and His Ways: Patterns of Life in Some Native Societies*. London, Odhams Press.

Blanchard, K. (1995). *The Anthropology of Sport*. Westport, Conn, Bergin &Garvey.

Blanchard, K. (2000). 'The Anthropology of Sport', in *Handbook of Sport Studies*, J. Coakley and E. Dunning, eds. London, Sage Publications.

Blum, D. (2002). *Love at Goon Park: Harry Harlow and the science of affection*. New York, Berkley Books.

Boas, O. V. and C. L. Boas (1973). *Xingu: The Indians, Their Myths*. New York, Farrar, Straus and Giroux.

Bogoraz-Tan, V. G. (1904). *The Chukchee: material culture, religion, social organization*. New York, G. E. Stechert & Co.

Borello, M. E. (2004). "'Mutual Aid' and 'Animal Dispersion': an historical analysis of alternatives to Darwin." *Perspectives in Biology and Medicine* **47**(1): 15-31.

Borello, M. E. (2005). "The rise, fall and resurrection of group selection." *Endeavor* **29**(1): 43-47.

Borg, J., B. Andree, H. Soderstrom and L. Farde (2003). "The Serotonin System and Spiritual Experiences." *American Journal of Psychiatry* **160**: 1965-1969.

Bourguignon, E. (1989). "Trance and Shamanism: What's in a Name?" *Journal of Psychoactive Drugs* **21**(1): 9-15.

Bower, H. (1989). "Beethoven's Creative Illness." *Australian and New Zealand Journal of Psychiatry* **23**: 111-116.

Bowlby, J. (1988). *A Secure Base*. New York, Basic Books.

Bowles, S. (2006). "Group competition, reproductive leveling, and the evolution of human altruism." *Science* **314**: 1569-1572.

Bowles, S. (2009). "Did Warfare Among Ancestral Hunter-Gatherers Affect the Evolution of Human Social Behaviors?" *Science* **324**(5932): 1293-1298.

Boyd, R. and P. J. Richerson (2002). "Group Beneficial Norms Can Spread Rapidly in a Structured Population." *Journal of Theoretical Biology* **215**(3): 287-296.

Boydell, J. and R. M. Murray (2003). Urbanization, migration and risk of schizophrenia. *The Epidemiology of Schizophrenia*. R. M. Murray, P. B. Jones, E. Susser, J. van Os and M. Cannon, Cambridge University Press.

Boyer, P. (2001). *Religion Explained: The Evolutionary Origins of Religious Thought*. New York, Basic Books.

Boyer, P. and C. Ramble (2001). "Cognitive templates for religious concepts: cross-cultural evidence for recall of counter-intuitive representations." *Cognitive Science* **25**: 535-564.

Bozikas, V. P., M. H. Kosmidis, M. Giannakou, D. Anezoulaki, P. Petrikis, K. Fokas and A. Karavatos (2007). "Humor appreciation deficit in schizophrenia: the relevance of basic neurocognitive functioning." *Journal of Nervous and Mental Disease* **195**(4): 325-331.

Bradbury, J. (2004). "Ancient footsteps in our genes: evolution and human disease." *Lancet* **363**(9413): 952-953.

Bradshaw, J. L. and D. M. Sheppard (2000). "The Neurodevelopmental Frontostriatal Disorders: Evolutionary Adaptiveness and Anomalous Lateralization." *Brain and Language* **73**: 297-320.

Brant, C. S., Ed. (1969). *Jim Whitewolf: The Life of a Kiowa Apache Indian*. New York, Dover.

Breed, M. D., G. E. Robinson and R. E. Page Jr (1990). "Division of labor during honey bee colony defense." *Behavioral Ecology and Sociobiology* **27**: 395-401.

Brenner, C. (1955/1974). *An Elementary Textbook of Psychoanalysis*. Garden City, New York, Anchor Books.

Bressloff, P. C., J. D. Cowan, M. Golubitsky, P. J. Thomas and M. C. Wiener (2002). "What geometric visual hallucinations tell us about the visual cortex." *Neural Computation* **14**(3): 473-491.

Brewer, M. B. (1999). "The psychology of prejudice: Ingroup Love or Outgroup Hate?" *Journal of Social issues* **55**(3): 429-444.

Brewerton, T. D. (1994). "Hyperreligiosity in psychotic disorders." *The Journal of Nervous and Mental Disease* **182**(5): 302-304.

Brewerton, T. D. (1997). "The phenomenology of psychosis associated with complex partial seizure disorder." *Annals of Clinical Psychiatry* **9**(1): 31-45.

Brewin, J., R. Cantwell, T. Dalkin, R. Fox, I. Medley, C. Glazebrook, R. Kwiecinski and G. Harrison (1997). "Incidence of schizophrenia in Nottingham. A comparison of two cohorts, 1978-80 and 1992-94." *The British Journal of Psychiatry* **171**: 140-144.

Brill, A. A., Ed. (1938). *The Basic Writings of Sigmund Freud*. New York, The Modern Library.

Broglio, A., M. De Stefani, F. Gurioli, P. Pallecchi, G. Giachi, T. Higham and F. Brock (2009). "The decoration of Fumane Cave in the picture of Aurignacian Art." *L'anthropologie* **113**: 753-761.

Brown, A. (2009). "Bridging the survival gap between Indigenous and non-Indigenous Australians: priorities for the road ahead." *Heart, Lung and Circulation* **18**(2): 96-100.

Brown, D. E. (1991). *Human Universals*. New York, McGraw-Hill.

Brown, P. (1997). Human origins and antiquity in Australia: An historical perspective. *History of Physical Anthropology: An Encyclopedia*. F. Spencer, Grand Publishing: 138-145.

Brugger, P. (2001). From Haunted Brain to Haunted Science: A cognitive neuroscience view of paranormal and pseudoscientific thought. *Hauntings and poltergeists: Multidisciplinary Perspectives*. J. Houran and R. Lange. Jefferson, NC, McFarland: 195-213.

Brugger, P., M. Regard, T. Landis, N. Cook, D. Krebs and J. Niederberger (1993). " 'Meaningful' patterns in visual noise: effects of lateral stimulation and the observer's belief in ESP." *Psychopathology* **26**(5-6): 261-265.

Brugger, P. and K. I. Taylor (2003). "ESP: Extrasensory Perception or Effect of Subjective Probability?" *Journal of Consciousness Studies* **10**: 221-246.

Brüne, M. (2009). On Shared Psychological Mechanisms of Religiousness and Delusional Beliefs. *The Biological Evolution of Religious Mind and Behavior*. W. Schiefenhovel. Berlin, Springer-Verlag.

Brüne, M. (2008). *Textbook of Evolutionary Psychiatry: The Origins of Psychopathology*. Oxford, Oxford University Press.

Brüne, M. (2004). "Schizophrenia — an evolutionary enigma?" *Neuroscience and Behavioral Reviews* **28**: 41-53.

Brüne, M. (2000). "Rethinking Karl Ludwig Kahlbaum's Contribution to Biological Psychiatry." *Australian and New Zealand Journal of Psychiatry* **34**(5): 873-874.

Buchanan, J. M. (2000). "Group Selection and Team Sports " *Journal of Bioeconomics* **2**: 1-7.

Buck, C., G. E. Hobbs, H. Simpson and J. M. Wanklin (1975). "Fertility of the Sibs of Schizophrenic Patients." *British Journal of Psychiatry* **127**: 235-239.

Buckley, P. (1981). "Mystical Experience and Schizophrenia." *Schizophrenia Bulletin* **7**(3): 516-521.

Burch, G. S., C. Pavelis, D. R. Hemsley and P. J. Corr (2006). "Schizotypy and creativity in visual artists." *British Journal of Psychology* **97**(Part 2): 177-190.

Burns, J. K. (2009). "Reconciling 'the new epidemiology' with an evolutionary genetic basis for schizophrenia." *Medical Hypotheses* **72**: 353-358.

Burns, K. B. (2004). "An evolutionary theory of schizophrenia: Cortical connectivity, metarepresentation, and the social brain." *Behavioral and Brain Sciences* **27**: 831-885.

Butcher, B. T. (1964). *My Friends, The New Guinea Headhunters*. Garden City, NY Doubleday & Co.

Cangas, A. J., L. A. Sass and M. Perez-Alvarez (2008). "From the Visions of Saint Teresa of Jesus to the Voices of Schizophrenia." *Philosophy, Psychiatry & Psychology* **15**(3): 239-250.

Cannon, M., P. B. Jones and R. M. Murray (2002). "Obstetric Complications and Schizophrenia: Historical and Meta-Analytic Review." *American Journal of Psychiatry* **159**: 1080-1092.

Cannon, M., R. Kendell, E. Susser and P. Jones (2003). Prenatal and perinatal risk factors for schizophrenia. *The Epidemiology of Schizophrenia*. R. M. Murray, P. B. Jones, E. Susser, J. van Os and M. Cannon, Cambridge University Press.

Cardno, A. G., E. J. Marshall, B. Coid, A. M. Macdonald, T. R. Ribchester, N. J. Davies, P. Venturi, L. A. Jones, S. W. Lewish, P. C. Sham, I. I. Gottesman, A. E. Farmer, P. McGuffin, A. M. Reveley and R. M. Murray (1999). "Heritability Estimates for Psychotic Disorders: The Maudsley Twin Psychosis Series." *Archives of General Psychiatry* **56**: 162-168.

Cardno, M. and R. M. Murray (2003). The 'Classical' Genetics epidemiology of schizophrenia. *The Epidemiology of Schizophrenia*. R. M. Murray, P. B. Jones, E. Susser, J. van Os and M. Cannon. Cambridge, UK, Cambridge University Press.

Carpenter, E. (1968). Witch-fear among the Aivilik Eskimos. *Eskimo of the Canadian Artic*. V. F. Valentine and F. G. Vallee. Toronto, McClelland and Stewart: 55-66.

Carson, S. H. (2011). "Creativity and psychopathology: A shared vulnerability model." *Canadian Journal of Psychiatry* **56**(3): 144-153.

Chapman, C. A., A. Gautier-Hion, J. F. Oates and D. A. Onderdonk (1999). African Primate Communities: Determinants of structure and threats to survival. *Primate Communities*. J. G. Fleagle, C. Janson and K. E. Reed. Cambridge, UK, Cambridge University Press.

Cheney, D. L. and R. M. Seyfarth (1990). *How Monkeys See The World*. Chicago, University of Chicago Press.

Choe, J. C. and B. J. Crespi, Eds. (1997). *The Evolution of Social Behavior in Insects and Arachnids*. Cambridge, UK, Cambridge University Press.

Choi, J. K. and S. Bowles (2007). "The Coevolution of Parochial Altruism and War." *Science* **318**(5850): 636-640.

Chui, H. T., B. K. Christensen, R. B. Zipursky, B. A. Richards, M. K. Hanratty, N. J. Kabani, D. J. Mikulis and D. K. Katzman (2008). "Cognitive Function and Brain Structure in Females with a History of Adolescent-Onset Anorexia Nervosa." *Pediatrics* **122**(2): e426-437.

Chun, R. (2002). "Bobby Fischer's Pathetic Endgame." *The Atlantic*.

Clegg, J. B. and D. J. Weatherall (1999). "Thalassemia and malaria: New insights into an old problem." *Proceedings of the Association of American Physicians* **111**(4): 278-282.

Combs, D. R. and D. L. Penn (2004). "The role of subclinical paranoia on social perception and behavior." *Schizophrenia Research* **69**: 93-104.

Combs, D. R., D. L. Penn and A. Fenigstein (2002). "Ethnic Differences in Subclinical Paranoia: An expansion of norms of the Paranoia Scale." *Cultural Diversity and Ethnic Minority Psychology* **8**(3): 248-256.

Coolidge, F. L., F. L. Davis and D. L. Segal (2007). "Understanding Madmen: A DSM-IV Assessment of Adolf Hitler." *Individual Differences Research* **5**(1): 30-43.

Cooper, J. R., F. E. Bloom and R. H. Roth (2003). *The Biochemical Basis of Neuropharmacology*. Oxford, Oxford University Press.

Corcoran, R., C. Cahill and C. D. Frith (1997). "The appreciation of visual jokes in people with schizophrenia: a study of 'mentalizing' ability." *Schizophrenia Research* **24**: 319-327.

Cox, D. and P. Cowling (1989). *Are You Normal?* London, Tower Press.

Craddock, N., M. C. O'Donovan and M. J. Owen (2005). "The genetics of schizophrenia and bipolar: dissecting psychosis." *Journal of Medical Genetics* **42**: 193-204.

Crespi, B., K. Summers and S. Dorus (2007). "Adapative evolution of genes underlying schizophrenia." *Proceedings. Biological sciences/ The Royal Society* **274**(1627): 2801-2810.

Cromwell, R. L. and J. M. Held (1969). "Alpha blocking latency and reaction time in schizophrenics and normals." *Perceptual and Motor Skills* **29**: 195-201.

Crow, T. J. (1997). "Is schizophrenia the price that Homo sapiens pay for language?" *Schizophrenia Research* **28**(2-3): 127-141.
Czaplicka, M. A. (1914). *Shamanism in Siberia*. Oxford, Forgotten Books.
d'Orsi, G. and P. Tinuper (2006). " 'I heard voices...': From semiology, a historical review, and a new hypothesis on the presumed epilepsy of Joan of Arc." *Epilepsy and Behavior* **6**: 152-157.
Dakin, S., P. Carlin and D. Hemsley (2005). "Weak suppression of visual context in chronic schizophrenia." *Current Biology* **15**(20): R822-824.
Dalgarno, P. (2007). "Subjective effects of Salvia divinorum." *Journal of Psychoactive Drugs* **39**(2): 143-149.
Darwin, C. (1859). *On the Origin of Species by Means of Natural Selection, or the Preservation of Favoured Races in the Struggle for Life*. London, John Murray.
Darwin, C. (1871). *The Descent of Man, and Selection in Relation to Sex*. London, John Murray.
Dawkins, R. (1976). *The Selfish Gene*. Oxford, Oxford University Press.
Dawkins, R. (2004). *The Ancestor's Tale: A Pilgrimage to the Dawn of Life*. London, Phoenix London.
Dawkins, R. (2006). *The God Delusion*. Boston, Houghton Mifflin.
Dawson, J. (1881). *Australian Aborigines: the language and customs of several tribes of Aborigines in the western district of Victoria, Australia*. Melbourne, George Robertson.
de Rios, M. D. (1984). *Hallucinogens: Cross-Cultural perspectives*. Albuquerque, University of New Mexico University Press.
De Waal, F. (1996). *Good Natured: The origins of right and wrong in humans and other animals*. Cambridge, MA, Harvard University Press.
De Waal, F., Ed. (2001). *Tree of Origin: What primate behavior can tell us about human social evolution*. Cambridge, MA, Harvard University Press.
Deacon, T. W. (1997). *The Symbolic Species: Co-evolution of Language and the brain*. New York, W.W. Norton & Co.
Dean, M., M. Carrington and S. J. O'Brien (2002). "Balanced Polymorphism Selected by Genetic Versus Infectious Human Disease." *Annual Review of Genomics and Human Genetics* **3**: 263-292.
Degenhardt, L. and W. Hall (2002). "Cannabis and Psychosis." *Current Psychiatry Reports* **4**(3): 191-196.
Dein, S. and R. Littlewood (2011). "Religion and Psychosis: A common evolutionary trajectory?" *Transcultural Psychiatry* **48**(3): 318-335.
DeLisi, L. E., K. U. Szulc, H. C. Bertisch, M. Majcher and K. Brown (2006). "Understanding structural changes in schizophrenia." *Dialogues in clinical neuroscience* **8**: 71078.
Dennett, D. C. (1995). *Darwin's Dangerous Idea: Evolution and the meaning of life*. New York, Touchstone.
Der, G., S. Gupta and R. M. Murray (1990). "Is schizophrenia disappearing?" *Lancet* **335**: 513-516.
DeRosse, P., A. Kaplan, K. E. Burdick, T. Lencz and A. K. Malhotra (2010). "Cannabis use disorders in schizophrenia: effects on cognition and symptoms." *Schizophrenia Research* **120**(1-3): 95-100.

Devereux, G. (1961). *Mohave Ethnopsychiatry and Suicide: The Psychiatric Knowledge and the Psychic Disturbances of an Indian Tribe*. Washington, DC., Government Printing Office.

Devereux, P. (1997). *The Long Trip: a prehistory of psychedelia*. New York, Penguin.

Devinsky, O. and G. Lai (2008). "Spirituality and religion in epilepsy." *Epilepsy & Behavior* **12**(4): 636-643.

Dewhurst, K. and A. W. Beard (2003). "Sudden religious conversions in temporal lobe epilepsy." *Epilepsy & Behavior* **4**: 78-87.

Doblin, R. (1991). "Pahnke's 'Good Friday experiment': A Long-Term Follow-Up and Methodological Critique." *The Journal of Transpersonal Psychology* **23**(1): 1-28.

Dobzhansky, T. (1973). "Nothing in Biology Makes Sense Except in the Light of Evolution" *The American Biology Teacher* **35**(March): 125-129.

Don, A., G. Schellenberg and B. Rourke (1999). "Music and language skills of children with Williams syndrome." *Child Neuropsychology* **5**: 154-170.

Donald, M. (1991). *Origins of the Modern Mind*. Cambridge, MA., Harvard University Press.

Dougherty, D. D., L. Baer, G. R. Cosgrove, E. H. Cassem, B. H. Price, A. A. Nierenberg, M. A. Jenike and S. L. Rauch (2002). "Prospective long-term follow-up of 44 patients who received cingulotomy for treatment-refractory obsessive-compulsive disorder." *American Journal of Psychiatry* **159**(2): 169-175.

Doughty, O. J. and D. J. Done (2009). "Is semantic memory impaired in schizophrenia? A systematic review and meta-analysis of 91 studies." *Cognitive Neuropsychiatry* **14**(6): 473-509.

Dowson, T. A. and A. L. Holliday (1989). "Zigzags and Eland: An Interpretation of an Idiosyncratic Combination." *The South African Archaeological Bulletin* **44**(149): 46-48.

Dowson, T. A. and M. Porr (2001). Special objects — special creatures: Shamanistic imagery and the Aurignacian art of south-west Germany. *The Archeology of Shamanism*. N. Price, Routledge.

Dreller, C. (1998). "Division of labor between scouts and recruits: genetic influence and mechanisms." *Behavioral Ecology and Sociobiology* **43**: 191-196.

Driscoll, C. A. and D. W. Macdonald (2009). "From wild animals to domestic pets, an evolutionary view of domestication." *Proceedings of the National Academy of Sciences USA* **106**(Suppl 1): 9971-9978.

Dugatkin, L. (1997). *Cooperation Among Animals: An evolutionary perspective*. New York, Oxford University press.

Dugatkin, L. (1999). *Cheating Monkeys and Citizen Bees: The nature of cooperation in animals and humans*. New York, The Free Press.

Dunbar, R. I. M. (1993). "Coevolution of neocortical size, group size and language in humans." *Behavioral and Brain Sciences* **16**: 681-735

Durkheim, É. (1912/2001). *The Elementary Forms of Religious Life*. Oxford, Oxford University Press.

Dykens, E. M., B. A. Rosner, T. Ly and J. Sagun (2005). "Music and anxiety in Williams syndrome: A harmonious or discordant relationship?" *American Journal of Mental Retardation* **110**(5): 346-358.

Eack, S. M., G. E. Hogarty, R. Y. Cho, K. M. Prasad, D. P. Greenwald, S. S. Hogarty and M. S. Keshavan (2010). "Neuroprotective effects of cognitive enhancement therapy against gray matter loss in early schizophrenia: results from a 2-year randomized controlled trial." *Archives of General Psychiatry* **67**(7): 674-682.

Eastwell, H. D. (1982). "Australian Aborigines." *Transcultural Psychiatric Research Review* **19**: 221-247.

Ebner, E., M. A. Broekema and B. Ritzler (1971). "Adaptation to Altered Visual-Proprioceptive Input in Normals and Schizophrenics." *Archives of General Psychiatry* **24**: 367-371.

Edgerton, R. B. (1966). "Conceptions of Psychosis in Four East African Societies." *American Anthropologist* **68**: 408-425.

Efferson, C., R. Lalive and E. Fehr (2008). "The Coevolution of Cultural Groups and Ingroup Favoritism." *Science* **321**(5897): 1844-1849.

Ehrenreich, B. (1997). *Blood Rites: origins and history of the passions of war*. New York, Henry Holt &Co.

Eliade, M. (1964). *Shamanism: Archaic Techniques of Ecstasy*. Princeton, Princeton University press.

Elkin, A. P. (1977). *Aboriginal Men of High Degree: Initiation and Sorcery in the World's Oldest Tradition*. Rochester, Vermont, Inner Traditions.

Ellett, L., B. Lopes and P. Chadwick (2003). "Paranoia in a Nonclincial Population of College Students." *Journal of Nervous and Mental Disease* **191**(7): 425-430.

Elliot, B., E. Joyce and S. Shorvon (2009). "Delusions, Illusions and Hallucinations in Epilepsy: 2. Complex phenomena and psychosis." *Epilepsy Research* **85**(2-3): 172-186.

Ellis, B. J., W. T. Boyce, J. Belsky, M. J. Bakermans-Kranenburg and M. H. van Ijzendoorn (2011). "Differential susceptibility to the environment: An evolutionary-neurodevelopmental theory." *Development and Psychopathology* **23**: 7-28.

Ellis, T. B. (2009). Natural Gazes, Non-natural agents: The biology of religion's ocular behaviours. *The biology of religious behaviour: The evolutionary orgins of faith and religion*. J. R. Feierman. Santa Barbara, CA, Praeger.

Endler, J. A. (1986). *Natural Selection in the Wild*. Princeton, NJ, Princeton University Press.

Epstein, E. and R. Guttman (1984). "Mate selection in Man: Evidence, theory, and outcome." *Social Biology* **31**(3-4): 243-278.

Esterberg, M. L. and M. T. Compton (2009). "The psychosis continuum and categorical versus dimensional diagnostic approaches." *Current Psychiatry Reports* **11**(3): 179-184.

Evans, F. W. (1858). *Ann Lee, The Founder Of The Shakers*. Mount Lebanon, New York, Charles Van Benthuvsen & Sons Printers.

Evans, K., J. McGrath and R. Milns (2003). "Searching for schizophrenia in ancient Greek and Roman literature: a systematic review." *Acta Psychiatrica Scandinavica* **107**(5): 323-330.

Fanciulli, M., E. Petretto and T. J. Aitman (2010). "Gene copy number variation and common human disease." *Clinical Genetics* **77**: 201-213.

Faulkes, C. G. and N. C. Bennet (2001). "Family Values: group dynamics and social control of reproduction in African mole-rats." *Trends in Ecology and Evolution* **16**(4): 184-190.

Ferm, V., Ed. (1965). *Ancient Religions*. New York, Citadel Press.

First, M. B. (2010). "Paradigm Shifts and the Development of the Diagnostic and Statistical Manual of Mental Disorders: Past Experiences and Future Aspirations." *Canadian Journal of Psychiatry* **55**(11): 692-700.

Fisher, J. E., A. Mohanty, J. D. Herrington, N. S. Koven, G. A. Miller and W. Heller (2004). "Neuropsychological evidence for dimensional schizotypy: Implications for Creativity and Psychopathology." *Journal of Research in Personality* **38**: 24-31.

Flint, J., A. V. S. Hill, D. K. Bowden, S. J. Oppenheimer, P. R. Sill, S. W. Serjeanston, J. Bana-Koiri, K. Bhatia, M. P. Alpers, A. J. Boyce, D. J. Weatherall and J. B. Clegg (1986). "High frequencies of α-thalassaemia are the result of natural selection by malaria." *Nature* **321**(19): 744-750.

Folsom, C. P. (1880). "A Lecture on Insanity — A Lecture Delivered before the Graduating Class of the Harvard Medical School." *Boston Medical and Surgical Journal* **102**: 481-484.

Fong, G. C., P. Kwan, A. C. Hui, C. H. Lui, J. K. Fong and V. Wong (2008). "An epidemiological study of epilepsy in Hong Kong SAR, China." *Seizure* **17**(5): 457-464.

Foster, L. (1993) "The Psychology of Religious Genius: Joseph Smith and the Origins of New Religious Movements." *Dialogue: A Journal of Mormon Thought* **26**(4): 1-22

Foucault, M. (1988). *Madness and Civilization: A History of Insanity in the Age of Reason*. New York, Vintage Books.

Francis, R. (2000). *Ann the Word: the story of Ann Lee, female messiah, mother of the Shakers, the women clothed with the sun*. New York, Arcade Publishing.

Fratkin, E. (2004). "The Laibon Diviner and Healer among Samburu Pastoralists of Kenya". *Divination and Healing*. M. Winkelman and P. M. Peek. Tucson, University of Arizona Press.

Frederiksen, S. (1968). Some Preliminaries on the Soul Complex in Eskimo Shamanistic Belief. *Eskimo of the Canadian Arctic*. V. F. Valentine and F. G. Vallee. Toronto, McClelland and Stewart.

Freeman, J. L., G. H. Perry, L. Feuk, R. Redon, S. A. McCarrol, D. M. Altshuler, H. Aburatuni, K. W. Jones, C. Tyler-Smith, M. E. Hurles, N. P. Carter, S. W. Scherer and C. Lee (2006). "Copy Number Variation: New insights in genome diversity." *Genome Research* **16**: 949-961.

Freemon, F. R. (1976). "A Differential Diagnosis of the Inspirational Spells of Muhammad the Prophet of Islam." *Epilepsia* **17**: 423-427.

Friedman, G. (2009) "Brian Wilson — A Powerful Interview." *Ability Magazine*.

Frischholz, E. J., L. S. Lipman, B. G. Braun and R. G. Sachs (1992). "Psychopathology, hypnotizability, and dissociation." *American Journal of Psychiatry* **149**(11): 1521-1525.

Frith, C. (2005a). "The Neural Basis of Hallucinations and Delusions." *Comptes Rendus Biologies* **328**(2): 169-175.

Frith, C. (2005b). "The Self in Action: Lessons from Delusions of Control." *Consciousness and Cognition* **14**: 752-770.

Fukuta, M. and E. Kirino (2004). "Event-related fMRI study of prefrontal and anterior cingulate cortex during response competition in schizophrenia." *International Congress Series* **1270**: 365-369.

Furst, P. T. (1976). *Hallucinogens and Culture*. Novato, CA, Chandler & Sharp.

Galton, D. (2009). "Did Darwin read Mendel?" *The Quarterly Journal of Medicine* **102**: 587-589.

Garber, P. A. and J. C. Bicca-Marques (2002). Evidence of predator sensitive foraging and traveling in single- and mixed-species tamarin troops. *Eat or be Eaten: predator sensitive foraging among primates*. L. E. Miller. Cambridge, UK, Cambridge University Press.

Garety, P. A. and D. Freeman (1999). "Cognitive approaches to delusions: A critical review of theories and evidence." *British Journal of Clinical Psychology* **38**: 113-154.

Garfinkel, Y. (2003). *Dancing at the Dawn of Agriculture*. Austin, University of Texas Press.

Garrett, C. (1935). *Origins of the Shakers*. Baltimore, Maryland, John Hopkins University Press.

Gately, I. (2001). *La Diva Nicotina: The Story of How Tobacco Seduced the World*. London, Simon & Schuster.

Gay, P. (1998). *Freud: a life for our time*. New York, W.W. Norton & Co.

Geary, D. C. (2004). *The Origin of Mind: Evolution of Brain, Cognition, and General Intelligence*. American Psychological Association.

Gemmell, N. J. and J. Slate (2006). "Heterozygote Advantage for Fecundity." *PLoS ONE* **1**(1): e125.

Getz, G. E., D. E. Fleck and S. M. Strkowski (2001). "Frequency and Severity of religious delusions in Christian patients with psychosis." *Psychiatry Research* **103**: 87-91.

Ghika, J. (2008). "Paleoneurology: Neurodegenerative diseases are age-related diseases of specific brain regions recently developed by homo sapiens " *Medical Hypotheses* **71**: 788-801.

Gladwell, M. (1996). "Conquering the Coma." *The New Yorker*.

Glicksohn, J. and H. Golan (2001). "Personality, cognitive style and assortative mating." *Personality and Individual Differences* **30**(7): 1199-1209.

Godfrey-Smith, P. (2003). *Theory and Reality: an introduction to the philosophy of science*. Chicago, The University of Chicago Press.

Goff, D. C., A. W. Brotman, D. Kindlon, M. Waites and E. Amico (1991). "The delusion of possession in chronically psychotic patients." *The Journal of Nervous and Mental Disease* **179**(9): 567-571.

Goldberg, T. E., J. D. Ragland, E. F. Torrey, J. M. Gold, L. B. Bigelow and D. R. Weinberger (1990). "Neuropsychological Assessment of Monozygotic Twins Discordant for Schizophrenia." *Archives of General Psychiatry* **47**(11): 1066-1072.

Goldner, E. M., L. Hsu, P. Waraich and J. M. Somers (2002). "Prevalence and Incidence Studies of Schizophrenic Disorders: A Systematic Review of the Literature." *Canadian Journal of Psychiatry* **47**: 833-843.

Goodall, J. (1986). *The Chimpanzees of Gombe: Patterns of Behavior*. Cambridge, MA, Belknap Press.

Goodall, J. (1990). *Through a Window: My Thirty Years with the Chimpanzees of Gombe*. Boston, Houghton Mifflin Co.

Goodnight, C. J. and L. Stevens (1997). "Experimental studies of group selection: What do they tell us about group selection in nature?" *The American Naturalist*. **150**(Suppl 1): s59-s79.

Gould, S. J. (2002). *The Structure of Evolutionary Theory*. Cambridge, MA, Belknap Press.

Goulding, A. (2004). *Mental Health Aspects of Paranormal and Psi Related Experiences*, Goteborg University.

Goulding, A. (2005). "Healthy schizotypy in a population of paranormal believers and experients." *Personality and Individual Differences* **38**: 1069-1083.

Greene, T. (2007). "The Kraepelinian Dichontomy: the twin pillars crumbling?" *History of Psychiatry* **18**(3): 361-379.

Griffiths, R. R., W. A. Richards, U. McCaan and R. Jesse (2006). "Psilocybin can occasion mystical-type experiences having substantial and sustained personal meaning and spiritual significance." *Psychopharmacology* **187**(3): 268-283.

Grim, J. A. (1983). *The Shaman:Patterns of Religious Healing Among the Ojibway Indians*. Norman, OK, University of Oklahoma Press.

Gross, M. R. (1991). "Evolution of alternative reproductive strategies: frequency-dependent sexual selection in male bluegill sunfish." *Philosophical Transactions of The Royal Society B* **332**: 59-66.

Grossman, D. (1995). *On Killing: The psychological cost of learning to kill in war and society*. Boston, Back Bay Books.

Grove, W. M. (1987). The reliability of psychiatric diagnosis. *Issues in diagnostic research*. C. G. Last and M. Hersen, Plenum Press: 99-119.

Guilaine, J. and J. Zammit (2001). *The Origins of War: violence in prehistory*. Malden, MA, Blackwell publishing.

Gur, R. E., M. S. Keshavan and S. M. Lawrie (2007). "Deconstructing Psychosis with Human Brain Imaging." *Schizophrenia Bulletin* **33**(4): 921-931.

Gutiérrez-Lobos, K., B. Schmid-Siegal, B. Bankier and H. Walter (2001). "Delusions in First-Admitted Patients: Gender, themes and diagnoses." *Psychopathology* **34**(1): 1-7.

Hafner, H. (2003a). "Gender differences in Schizophrenia." *Psychoneuroendocrinology* **28**(Suppl 2): 17-54.

Hafner, H. and W. an der Heiden (1997). "Epidemiology of Schizophrenia." *Canadian Journal of Psychiatry* **42**: 139-151.

Hakim, A. B. (1967/2001). *Historical Introduction to Philosophy* Upper Saddle River, NJ, Prentice Hall.

Hallgrimsson, B. and B. K. Hall, Eds. (2005). *Variation: A central concept in biology*. Burlington, MA, Elsevier Academic Press.

Hamilton, W. D. (1971). "Geometry for a selfish herd." *Journal of Theoretical Biology* **31**: 295-311.

Harner, M. (1980). *The Way of the Shaman*. New York, Harper & Row.

Harner, M. J., Ed. (1973). *Hallucinogens and Shamanism*. London, Oxford University Press.

Harris, P. M. G. (2003). *The History of Human Populations: Vol II — Migration, urbanization, and structural changes*. Westport, Connecticut, Praeger.

Harrison, P. J. (1999). "The Neuropathology of Schizophrenia: a critical review of the data and their interpretation." *Brain* **122**(4): 593-624.

Harrison, P. J. and D. R. Weinberger (2005). "Schizophrenia genes, gene expression, and neuropathology: on the matter of their convergence." *Molecular Psychiatry* **10**: 40-68.

Hasler, F., U. Grimberg, M. A. Benz, T. Huber and F. X. Vollenweider (2004). "Acute psychological and physiological effects of psilocybin in Healthy humans: a double-blind, placebo-controlled dose-effect study." *Psychopharmacology* **172**(2): 424-433.

Haukka, J., J. Suvisaari and J. Lonnqvist (2003). "Fertility of patients with schizophrenia, their siblings, and the general population: A cohort study from 1950-1959 in Finland." *American Journal of Psychiatry* **160**(3): 460-463.

Haverkamp, F., P. Propping and T. Hilger (1982). "Is there an increase of reproductive rates in schizophrenia?" *Archiv fur Psychiatrie und Nervenkrankheiten* **232**: 439-450.

Healey, D. and J. J. Rucklidge (2006). "An investigation into the relationship among ADHD symptomatology, creativity, and neuropsychological functioning in children." *Child Neuropsychology* **6**: 421-438.

Healy, D. (2002). *The Creation of Psychopharmacology*. Cambridge, MA, Harvard University Press.

Heider, K. G. (1970). *The Dugum Dani: A Papuan Culture in the Highlands of West New Guinea*. Chicago, Aldine Publishing.

Heinrichs, R. (2003). "Historical origins of schizophrenia: Two early madmen and their illness." *Journal of the History of the Behavioral Sciences* **39**(4): 349-363.

Heinrichs, R. W. and K. K. Zakzanis (1998). "Neurocognitive Deficit in Schizophrenia: A Quantitative Review of the Evidence." *Neuropsychology* **12**(3): 426-445.

Helling, I., A. Ohman and C. M. Hultman (2003). "School achievements and schizophrenia: a case-control study." *Acta Psychiatrica Scandinavica* **108**: 381-386.

Herodotus (1998). *The Histories*. Oxford, Oxford University Press.

Hershman, D. J. and J. Lieb (1998). *Manic Depression and Creativity*. New York, Prometheus Books.

Hickling, F. W. (1999). "A Jamaican Psychiatrist Evaluates Diagnoses at a London Psychiatric Hospital." *British Journal of Psychiatry* **175**: 283-285.

Ho, B. C., N. C. Andreasen, S. Ziebell, R. Pierson and V. Magnotta (2011). "Long-term Antipsychotic Treatment and Brain Volumes: A Longitudinal Study of First-Episode Schizophrenia." *Archives of General Psychiatry* **68**(2): 128-137.

Hoenig, J. (1984). "Schneider's first rank symptoms and the tabulators." *Comprehensive Psychiatry* **25**(1): 77-87.

Hoff, A. L. and W. S. Kremen (2003). "Neuropsychology in Schizophrenia: An update." *Current Opinion in Psychiatry* **16**(2): 149-155.

Hoffman, R. E., M. Varanko, J. Gilmore and A. L. Mishara (2008). "Experiential features used by patients with schizophrenia to differentiate 'voices' from ordinary verbal thought." *Psychological Reports* **38**(8): 1167-1176.

Hoistad, M., D. Segal, N. Takahashi, T. Sakurai, J. D. Buxbaum and P. R. Hof (2009). "Linking white and grey matter in schizophrenia: Oligodendrocyte and neuron pathology in the prefrontal cortex." *Frontiers in Neuroanatomy* **3**: 9.

Hölldobler, B. and E. O. Wilson (1994). *Journey to the Ants: a story of scientific exploration.* Cambridge, MA, Belknap Press.

Hood, R. and R. Morris (1981). "Sensory Isolation and the Differential report of visual imagery in intrinsic and extrinsic subjects." *Journal for the Scientific Study of Religion* **20**: 261-273.

Hopyan, T., M. Dennis, R. Weksberg and C. Cytrynbaum (2001). "Music Skills and the Expressive Interpretation of Music in Children with Williams-Beuren Syndrome: Pitch, Rhythm, Melodic Imagery, Phrasing, and Musical Affect. ." *Child Neuropsychology* **7**(1): 42-53.

Hor, K. and M. Taylor (2010). "Suicide and Schizophrenia: a systematic review of rates and risk factors." *Journal of Psychopharmacology* **24**(4 Suppl.): 81-90.

Horgan, J. (1997). *The End of Science: Facing the Limits of knowledge in the Twilight of the Scientific Age.* New York, Broadway Books.

Hori, M. (1993). "Frequency-Dependent Natural Selection in the Handedness of Scale-Eating Cichlid Fish." *Science* **260**(5105): 216-219.

Horrobin, D. F. (1998). "Schizophrenia: the illness that made us human." *Medical Hypotheses* **50**: 269-288.

Hudjashov, G., T. Kivisild, P. A. Underhill, P. Endicott, J. J. Sanchez, A. A. Lin, P. Shen, P. Oefner, C. Renfrew, R. Villems and P. Forster (2007). "Revealing the prehistoric settlement of Australia by Y chromosome and mtDNA analysis." *Proceedings of the National Academy of Sciences USA* **104**(21): 8726-8730.

Huguelet, P. and N. Perroud (2005). "Wolfgang Amadeus Mozart's Psychopathology in Light of the Current Conceptualization of Psychiatric Disorders." *Psychiatry* **68**(2): 130-139.

Hunsberger, B. E. and B. Altemeyer (2006). *Atheists: A groudbreaking study of America's Nonbelievers.* New York, Prometheus Books.

Hunt, M. (1993/2007). *The Story of Psychology.* New York, Anchor Books.

Hur, Y. M. (2003). "Assortative mating for personality traits, educational level, religious affiliation, height, weight, and body mass index in parents of a Korean twin sample." *Twin Research* **6**(6): 467-470.

Huxley, F. (1957). *Affable Savages: An Anthropologist Among the Urubu Indians of Brazil.* New York, Viking Press.

Huxley, J., E. Mayr, H. Osmond and A. Hoffer (1964). "Schizophrenia as a genetic morphism." *Nature* **204**: 220-221.

Hyde, T. M. and D. R. Weinberger (1997). "Seizures and Schizophrenia " *Schizophrenia Bulletin* **23**(4): 611-622.

Iacoboni, M., M. D. Lieberman, B. J. Knowlton, I. Molnar-Szakacs, M. Moritz, J. Throop and A. P. Fiske (2004). "Watching social interactions produces dorsomedial prefrontal and medial parietal BOLD fMRI signal increases compared to a resting baseline." *Neuroimage* **21**: 1167-1173.

Iida, J. (1989). "The current situation in regard to the delusion of possession in Japan." *The Japanese journal of psychiatry and neurology* **43**(1): 19-27.

Ikeda, M., B. Aleksic, G. Kirov, Y. Kinoshita, Y. Yamanouchi, T. Kitajima, K. Kawashima, T. Okochi, T. Kishi, I. Zaharieva, M. J. Owen, M. C. O'Donovan, N. Ozaki and N. Iwata (2010). "Copy Number Variation in Schizophrenia in the Japanese Population." *Biological Psychiatry* **67**(3): 283-286.

Ingle, H. L. (1994). *First Among Friends: George Fox and the creation of Quakerism.* Oxford, Oxford University Press.

Ingoldsby, B. B. (2006). Religious Utopias and Family Structure. *Families in Global and Multicultural Perspective.* B. B. Ingoldsby and S. D. Smith. London, Sage Publications.

Ireland, W. W. (1883). "On the Character and Hallucinations of Joan of Arc." *Journal of Mental Science* **28**: 483-492.

Iritani, S. (2007). "Neuropathology of Schizophrenia: A mini review." *Neuropathology* **27**: 604-608.

Irwin, H. J. (1993). "Belief in the Paranormal: A review of the empirical literature." *The Journal of the American Society for Psychical Research* **87**(1): 1-39.

Irwin, L. (1994). *The Dream Seekers: Native American Visionary Traditions of the Great Plains.* Norman, OK, University of Oklahoma Press.

Jablensky, A. (2003). *The Epidemiological Horizon In Schizophrenia.* Massachusetts, Blackwell Science.

Jablensky, A. and S. W. Cole (1997). "Is the earlier age at onset of schizophrenia in males a confounded finding?" *British Journal of Psychiatry* **170**: 234-240.

Jablensky, A., N. Sartorius, G. Ernberg, M. Anker, A. Korten, J. E. Cooper, R. Day and A. Bertelsen (1992). "Schizophrenia: manifestations, incidence and course in different cultures. A World Health Organization Ten-Country Study" *Psychological Medicine Monograph Supplement* **20**: 1-97.

Jablonka, E. and M. J. Lamb (2006). *Evolution in Four Dimensions: Genetic, Epigenetic, Behavioral, and Symbolic Variation in the History of Life.* Cambridge, MA, MIT Press.

James, F. E. (1992). "Insulin Treatment in Psychiatry." *History of Psychiatry* **3**(10): 221-235.

James, W. (1902). *The Varieties of Religious Experience: A study in Human Nature.* Cambridge, MA, Harvard University Press.

Jamieson, G. A. and J. H. Gruzelier (2006). "Hypnotic susceptibility is positively related to a subset of schizotypy items." *Contemporary Hypnosis* **18**(1): 32-37.

Jamison, K. R. (1989). "Mood Disorders and Patterns of creativity in British writers and artists." *Psychiatry* **52**(2): 125-134.

Jamison, K. R. (1995). "Manic-depressive illness and creativity." *Scientific American* **272**(2): 62-67.

Janson, C. H. and M. L. Goldsmith (1995). "Predicting group size in primates: foraging costs and predation risks." *Behavioral Ecology* **6**(3): 326-336.

Jaynes, J. (1976). *The Origin of Consciousness in the Breakdown of the Bicameral Mind.* Boston, Houghton Mifflin Co.

Jeste, D. V., R. Del Carmen, J. B. Lohr and R. J. Wyatt (1985). "Did schizophrenia exist before the eighteenth century?" *Comprehensive Psychiatry* **26**(6): 493-503.

Jeste, D. V., K. A. Harless and B. W. Palmer (2000). "Chronic Late-Onset Schizophrenia-Like Psychosis That Remitted: Revisiting Newton's Psychosis?" *American Journal of Psychiatry* **157**(3): 444-449.

Johns, L. C., J. Y. Nazroo, P. Bebbington and E. Kuipers (2002). "Occurrence of hallucinatory experiences in a community sample and ethnic variations." *British Journal of Psychiatry* **180**: 174-178.

Johns, L. C. and J. van Os (2001). "The Continuity of Psychotic Experiences in the General Population." *Clinical Psychology Review* **21**(8): 1125-1141.

Johnson, D. E. and G. D. Shean (1993). "Word Associations and Schizophrenic Symptoms " *Journal of Psychiatric Research* **27**(1): 69-77.

Johnson, J. (1994). "Henry Maudsley on Swedenborg's Messianic Psychosis." *British Journal of Psychiatry* **165**: 690-691.

Johnson, J. A., S. U. Vermeulen, E. L. Toth, B. R. Hemmelgarn, K. Ralph-Campbell, G. Hugel, M. King and L. Crowshoe (2009). "Increasing incidence and prevalence of diabetes among the Status Aboriginal population in urban and rural Alberta, 1995-2006." *Canadian Journal of Public Health* **100**(3): 231-236.

Jones, I. H. and D. J. Horne (1973). "Psychiatric Disorders among Aborigines of the Australian Western Desert." *Social Science & Medicine* **7**: 219-228.

Judd, T. M. and P. W. Sherman (1996). "Naked mole-rats recruit colony mates to food sources." *Animal Behaviour* **52**: 957-969.

Junod, H. A. (1962). *The Life of a South African Tribe.* New Hyde Park, NY., University Books.

Kallmann, F. J. (1946). "The Genetic Theory of Schizophrenia: An Analysis of 691 Schizophrenic Twin Index Families." *American Journal of Psychiatry* **151**(Suppl 6): 189-198.

Kanner, A. M. (2000). "Psychosis of Epilepsy: A Neurologist's Perspective." *Epilepsy Behavior* **1**(4): 219-227.

Kapogiannis, D., A. K. Barbey, M. Su, F. Krueger and J. Grafman (2009a). "Neuroanatomical Variability of Religiosity." *PLoS One* **4**(9): e7180.

Kapogiannis, D., A. K. Barbey, M. Su, G. Zamboni, F. Krueger and J. Grafman (2009b). "Cognitive and neural foundations of religious belief." *Proceedings of the National Academy of Sciences USA* **106**(12): 4876-4881.

Karlsson, J. L. (1974). "Inheritance of schizophrenia." *Acta Psychiatrica Scandinavica* **274**: 77-88.

Karlsson, J. L. (1984). "Creative intelligence in relatives of mental patients." *Heriditas* **100**(1): 83-86.

Karlsson, J. L. (1999). "Relation of mathematical ability to psychosis in Iceland." *Clinical Genetics* **56**(6): 447-449.

Karlsson, J. L. (2001). "Mental abilities of male relatives of psychotic patients." *Acta Psychiatrica Scandinavica* **104**(6): 466-468.

Kauffman, C., H. Grunebaum, B. Cohler and E. Gamer (1979). "Superkids: Competent Children of Psychotic Mothers." *American Journal of Psychiatry* **136**(11): 1398-1402.

Keefe, J. A. and P. A. Magaro (1980). "Creativity and Schizophrenia: An Equivalence of Cognitive Processing." *Journal of Abnormal Psychology* **89**(3): 390-398.

Keeley, L. H. (1996). *War Before Civilization: The myth of the peaceful savage*. New York, Oxford University Press.

Keller, M. C. and G. Miller (2006). "Resolving the paradox of common, harmful, heritable mental disorders: Which evolutionary genetic models work best?" *Brain and Behavioral Sciences* **29**: 385-452.

Kendell, R. E., D. E. Malcolm and W. Adams (1993). "The problem of detecting changes in the incidence of schizophrenia." *The British Journal of Psychiatry* **162**: 212-218.

Kendler, K. S., A. M. Gruenberg and D. K. Kinney (1994). "Independent diagnoses of adoptees and relatives as defined by DSM-III in the provincial and national samples of the Danish adoption study of schizophrenia." *Archives of General Psychiatry* **51**: 456-468.

Kendler, K. S., A. M. Gruneberg and J. S. Strauss (1981). "An independent analysis of the Copenhagen sample of the Danish Adoption Study of Schizophrenia: II. The relationship between schizotypal personality disorder and schizophrenia." *Archives of General Psychiatry* **38**(9): 982-984.

Kensinger, K. M. (1973). Banisteriopsis Usage Among the Peruvian Cashinahua. *Hallucinogens and Shamanism*. M. J. Harner. Oxford, Oxford University Press.

Kent, G. H. and A. J. Rosanoff (1910). "A study of association in insanity." *American Journal of Insanity* **66**: 37-47.

Keshavan, M. S., R. Tandon, N. N. Boutros and H. A. Nasrallah (2008). "Schziophrenia, 'just the facts': What we know in 2008 — Part 3 Neurobiology." *Schizophrenia Research* **106**: 89-107.

Ketter, T. A., P. O. Wang, O. V. Becker, C. Nowakowska and Y. Yang (2004). "Psychotic bipolar disorders: dimensionally similar to or categorically different from schizophrenia?" *Journal of Psychiatric Research* **38**: 47-61.

Kety, S. S., P. H. Wender, B. Jacobsen, L. J. Ingraham, L. Jansson, B. Faber and D. K. Kinney (1994). "Mental illness in the biological and adoptive relatives of schizophrenic adoptees. Replication of the Copenhagen study in the rest of Denmark." *Archives of General Psychiatry* **51**(6): 442-455.

Keynes, M. (1980). "Sir Isaac Newton and his Madness of 1692-93." *Lancet* **1**(8167): 529-530.

Khaitovich, P., H. E. Lockstone, M. T. Wayland, T. M. Tsang, S. D. Jayatilaka, A. J. Guo, J. Zhou, M. Somel, L. W. Harris, E. Holmes, S. Pääbo and S. Bahn (2008). "Metabolic changes in schizophrenia and human brain evolution." *Genome Biology* **9**(8): R124.

Kim, C. H., J. W. Chang, M. S. Koo, J. W. Kim, H. S. Suh, I. H. Park and H. S. Lee (2003a). "Anterior cingulotomy for refractory obsessive-compulsive disorder." *Acta Psychiatrica Scandinavica* **107**(4): 283-290.

Kim, K. I., Z. Jiang, X. Cui, L. Lin, J. J. Kang, K. K. Park, E. K. Chung and C. K. Kim (1993). "Schizophrenic delusions among Koreans, Korean-Chinese and Chinese: a transcultural study." *International Journal of Social Psychiatry* **39**(3): 190-199.

Kim, K. Y., R. H. Schumacher, E. Hunsche, A. I. Wertheimer and S. X. Kong (2003b). "A Literature Review of the Epidemiology and Treatment of Acute Gout." *Clinical Therapeutics* **25**(6): 1593-1617.

Kimhy, D., R. Goetz, S. Wale, C. Corcoran and S. Malaspina (2005). "Delusions in Individuals with schizophrenia: Factor Structure, Clinical Correlates, and Putative Neurobiology " *Psychopathology* **38**: 338-344.

King, L. A., C. M. Burton, J. A. Hicks and S. M. Drigotas (2007). "Ghosts, UFOs, and Magic: Positive Affect and the Experiential System." *Journal of Personality and Social Psychology* **95**(5): 905-919.

Kingdon, J. (1993). *Self-Made Man: Human Evolution from Eden to Extinction?* New York, John Wiley & Sons.

Kirov, G., D. Grozeva, N. Norton, D. Ivanov, K. K. Mantripragada, P. Holmans, ISCt. WTCC Consortium)., N. Craddock, M. J. Owen and M. C. O'Donovan (2009). "Support for the involvement of large copy number variants in the pathogenesis of schizophrenia " *Human Molecular Genetics* **18**(8): 1497-1503.

Kishimoto, M., H. Ujike, Y. Motohashi, Y. Tanaka, Y. Okahisa, T. Kotaka, M. Harano, T. Inada, M. Yamada, T. Komiyama, T. Hori, Y. Sekine, N. Iwata, I. Sora, M. Iyo, N. Ozaki and S. Kuroda (2008). "The dysbindin gene (DTNBP1) is associated with methamphetamine psychosis." *Biological Psychiatry* **63**(2): 191-196.

Kitade, O., M. Hoshi, S. Odaira, A. Asano, M. Shimizu, Y. Hayashi and N. Lo (2011). "Evidence for genetically influenced caste determination in phylogenetically diverse species of the termite genus *Reticulitermes*." *Biology Letters* **7**: 257-260.

Klein, R. G. and B. Edgar (2002). *The Dawn of Human Culture.* New York, John Wiley & Sons.

Konopaske, G. T., R. A. Sweet, Q. Wu, A. Sampson and D. A. Lewis (2006). "Regional specificity of chandelier neuron axon terminal alterations in schizophrenia." *Neuroscience* **138**(1): 189-196.

Kosten, T. R. and D. M. Ziedonis (1997). "Substance abuse and Schizophrenia: Editors' introduction." *Schizophrenia Bulletin* **23**(2): 181-186.

Kramer, E. and E. P. Brennan (1964). "Hypnotic susceptibility of schizophrenic patients." *Journal of Abnormal and Social Psychology* **69**(6): 657-659.

Krippner, S. C. (2002). "Conflicting Perspectives on Shamans and Shamanism: Points and Counterpoints." *American Psychologist* **57**: 962-977.

Kroll, J. (1973). "A Reappraisal of Psychiatry in the Middle Ages." *Archives of General Psychiatry* **29**: 276-283.

Kroll, J. and B. Bachrach (1982). "Medieval visions and Contemporary hallucinations." *Psychological Medicine* **12**: 709-721.

Kropotkin, P. (1902/1972). *Mutual Aid: a factor in evolution.* New York, New York University Press.

Krysanki, V. and F. R. Ferraro (2007). "Creativity, unique designs, and schizotypy in a nonclinical sample." *Psychological Reports* **101**(1): 273-274.

Kua, E. H., P. H. Chew and S. M. Ko (1993). "Spirit possession and healing among Chinese psychiatric patients." *Acta Psychiatrica Scandinavica* **88**(6): 447-450.

Kubicki, M., R. McCarley, C. F. Westin, H. J. Park, S. Maier, R. Kikinis, F. A. Jolesz and M. E. Shenton (2007). "A review of diffusion tensor imaging studies in schizophrenia " *Journal of Psychiatric Research* **41**: 15-30.

Kuhn, T. S. (1962). *The Structure of Scientific Revolutions*. Chicago, The University of Chicago Press.

Kulhara, P., A. Avasthi and A. Sharma (2000). "Magico-Religious Beliefs in Schizophrenia: A Study from North India." *Psychopathology* **33**(2): 62-68.

Kumari, V. and M. Cooke (2006). "Use of magnetic resonance imaging in tracking the course and treatment of schizophrenia." *Expert Review of Neurotherapeutics* **6**(7): 1005-1016.

Kuperberg, G. R., T. Deckersbach, D. J. Holt, D. Goff and C. West (2007). "Increased Temporal and Prefrontal Activity in Response to Semantic Associations in Schizophrenia." *Archives of General Psychiatry* **64**(2): 138-151.

Kurtz, M. M. (2005). "Neurocognitive impairment across the lifespan in schizophrenia: an update." *Schizophrenia Research* **74**(15-26).

Kurtz, P. (1986). *The Transcendental Temptation: A Critique of Religion and the Paranormal*. Amherst, New York, Prometheus Books.

La Barre, W. (1970). *The Ghost Dance: The Origins of Religion*. New York, Dell Publishing.

La Barre, W. (1972). Hallucinogens and the Shamanic Origins of Religion. *Flesh of the Gods: The Ritual use of Hallucinogens*. P. T. Furst. Long Grove, Illinois, Waveland Press.

Laale, H. W. (2007). *Once They Were Brave, The Men of Miletus*. Bloomington, Indiana, AuthorHouse.

Lam, C. W. and B. Berrios (1992). "Psychological concepts and psychiatric symptomatology in some ancient Chinese medical texts." *History of Psychiatry* **3**: 117-128.

Lange-Eichbaum, W. (1932). *The Problem of Genius*. New York, MacMillan Publishers.

Langness, L. L. (1965). "Hysterical Psychosis in the New Guinea Highlands: A Bena Bena Example." *Psychiatry* **28**: 258-277.

Lanternari, V. (1963). *The religions of the oppressed*. New York, Mentor Books.

Lapierre, Y. D. (1994). "Schizophrenia and Manic-Depression: Separate illnesses or a continuum?" *Canadian Journal of Psychiatry* **39**(9, Suppl. 2): S59-S64.

Laubscher, B. F. J. (1937). *Sex, Custom and Psychopathology:A study of South African Pagan Natives*. London, Routledge & Sons.

Lauronen, E., J. Veijola, I. Isohanni, P. B. Jones, P. Nieminen and M. Isohanni (2004). "Links between creativity and mental disorder." *Psychiatry* **67**(1): 81-98.

Lautenschlager, N. T. and A. F. Kurz (2010). Neurodegenerative disorders (Alzheimer's disease, fronto-temporal dementia) and schizophrenic-like psychosis. *Secondary Schizophrenia*. P. S. Sachdev and M. S. Keshaven. Cambridge, Cambridge University Press.

Lawrence, E. and E. Peters (2004). "Reasoning in Believers in the Paranormal." *The Journal of Nervous and Mental Disease* **192**: 727-733.

Lawrie, S. M., J. Hall, A. M. McIntosh, D. G. Owens, E. C. Johnstone (2010). "The 'continuum of psychosis': scientifically unproven and clinically impractical." *British Journal of Psychiatry* **197**(6): 423-425.

Ledgerwood, L. G., P. W. Ewald and G. M. Cochran (2003). "Genes, Germs and Schizophrenia: an evolutionary perspective." *Perspectives in Biology and Medicine* **46**(3): 317-348.

Lenhoff, H., O. Perales and G. Hickok (2001). "Absolute pitch in Williams syndrome." *Music Perception* **18**(3): 491-503.

Leonard, B. E. (1997). *Fundamentals of Psychopharmacology*. Chichester, UK. John Wiley and Sons.

Leonard, S., C. Breese, C. Adams, K. Benhammou, J. Gault, K. Stevens, M. Lee, L. Adler, A. Olincy, R. Ross and R. Freedman (2000). "Smoking and Schizophrenia: abnormal nicotinic receptor expression." *European Journal of Pharmacology* **393**: 237-242.

Leudar, I. and P. Thomas (2008). *Voices of Reason, Voices of Insanity: Studies of Verbal Hallucinations*. London, Routledge.

Leung, A. and P. Chue (2000). "Sex Differences in Schizophrenia: a review of the literature." *Acta Psychiatrica Scandinavica* **401**: 3-38.

Levitin, D. J. (2006). *This is Your Brain on Music*. New York, Dutton Books.

Levitin, D. J. (2008). *The World in Six Songs: how the musical brain created human nature*. Toronto, Viking Canada.

Lewis, I. M. (2003). *Ecstatic Religion: A study of Shamanism and Spirit Possession*. London, Routledge.

Lewis, J. R., Ed. (2001). *Odd Gods: New Religions and the Cult Controversy*. Amherst, NY Prometheus Books.

Lewis-Williams, J. D. (1997). Harnessing the Brain: Vision and Shamanism in Upper Paleolithic Western Europe. *Beyond Art: Pleistocene Image and Symbol*. M. Conkey, O. Soffer, D. Stratmann and N. G. Jablonski. San Francisco, University of California Press.

Lewis-Williams, J. D. (2001). Brainstorming Images: Neuropsychology and Rock Art Research. *Handbook of Rock art Research*. D. S. Whitley. Walnut Creek, CA, Altamira Press.

Lindberg, D. C. (2007). *The Beginnings of Western Science: The European Scientific Tradition in Philosophical, Religious, and Institutional Context, Prehistory to A.D. 450*. Chiacgo, University of Chicago Press.

Linde, P. R. (2002). *Of Spirits and Madness: An American Psychiatrist in Africa*. New York, McGraw-Hill.

Lindeman, M. and K. Aarnio (2006). "Paranormal Beliefs: Their Dimensionality and Correlates." *European Journal of Personality* **20**: 585-602.

Lindholm, C. (1990). *Charisma*. Cambridge, MA, Basil Blackwell.

Linscott, R. J. and J. van Os (2010). "Systematic reviews of categorical versus continuum models in psychosis: evidence for discontinous subpopulations underlying a psychometric continuum. Implications for DSM-V, DSM-VI and DSM-VII." *Annual Review of Clinical Psychology* **6**: 391-419.

Lipska, B. K. (2004). "Using animal models to test a neurodevelopmental hypothesis of schizophrenia." *Journal of Psychiatry & Neuroscience* **29**(4): 282-286.

Lipska, B. K. and D. R. Weinberger (2000). "To model a psychiatric disorder in animals: schizophrenia as a reality test." *Neuropsychopharmacology* **23**: 223-239.

Loberg, E. M. and K. Hugdahl (2009). "Cannabis Use and Cognition in Schizophrenia." *Frontiers in Human Neuroscience* **3**(53): 1-8.

Loeb, E. M. (1924). "The Shaman of Niue." *American Anthropologist* **26**(3): 393-402.

Lombardo, M. P. (2012). "On the evolution of sport." *Evolutionary Psychology* **10**(1): 1-28.

Loranger, A. W. (1981). "Genetic Independence of Manic-Depression and Schizophrenia." *Acta Psychiatrica Scandinavica* **63**: 444-452.

Lucas, R. H. and R. J. Barret (1995). "Interpreting Culture and Psychopathology: Primitivist Themes in Cross-cultural Debate." *Culture, Medicine and Psychiatry* **19**: 287-326.

Ludwig, A. M. (1992). "Creative Achievement and Psychopathology: Comparison among Professions." *American Journal of Psychotherapy* **46**(3): 330-356.

Ludwig, A. M. (1995). *The Price of Greatness: Resolving the Creativity and Madness Controversy*. New York, Guilford Press.

Luo, S. and E. C. Klohnen (2005). "Assortative mating and marital quality in newlyweds: A couple-centered approach." *Journal of Personality and Social Psychology* **88**(2): 304-326.

Lyell, C. (1830/1997). *Principles of Geology*. London, Penguin Books.

Lyon, W. S. (2004). "Divination in North American Indian Shamanic Healing". *Divination and Healing*. M. Winkelman and P. M. Peek. Tucson, University of Arizona Press.

MacCabe, J. H., I. Koupil and D. A. Leon (2009). "Lifetime reproductive output over two generations in patients with psychosis and their unaffected siblings: the Uppsala 1915-1929 Birth Cohort Multigenerational Study." *Psychological Medicine* **39**(10): 1667-1676.

MacNeill, A. D. (2004). "The Capacity for Religious Experience is an Evolutionary Adaptation to Warfare." *Evolution and Cognition* **10**(1): 43-60.

Maghazaji, H. I. (1974). "Psychiatric Aspects of Methylmercury poisoning " *Journal of Neurology, Neurosurgery & Psychiatry* **37**: 954-958.

Maher, B. A. (1992). "Delusions: Contemporary etiological hypotheses." *Psychiatric Annals* **22**: 260-268.

Malik, S. B., M. Ahmed, A. Bashir and T. M. Choudhry (1990). "Schneider's first-rank symptoms of schizophrenia: prevalence and diagnostic use. A study from Pakistan." *The British Journal of Psychiatry* **156**: 109-111.

Malthus, T. R. (1798). *An essay on the principle of population as it affects the future improvement of society*. London, J. Johnson.

Malur, C., M. Fink and A. Francis (2000). "Can Delirium Relieve Psychosis?" *Comprehensive Psychiatry* **41**(6): 450-453.

Mann, J., R. C. Connor, P. L. Tyack and H. Whitehead, Eds. (2000). *Cetacean Societies: Field studies of dolphins and whales*. Chicago, The University of Chicago Press.

Manoach, D. S. (2003). "Prefrontal cortex dysfunction during working memory performance in schizophrenia: reconciling discrepant findings." *Schizophrenia Research* **60**(2-3): 285-298.

Manschreck, T. C., B. A. Maher, J. L. Milavetz, D. Ames, C. C. Weisstein and M. L. Schneyer (1988). "Semantic Priming in Thought Disordered Schizophrenic Patients " *Schizophrenia Research* **1**(1): 61-66.

Marcotte, E. R., D. M. Pearson and L. K. Srivastava (2001). "Animal Models of Schizophrenia: a critical review." *Journal of Psychiatry & Neuroscience* **26**(5): 395-410.

Martin, L. F., M. Hall, R. G. Ross, G. Zerbe, R. Freedman and A. Olincy (2007). "Physiology of Schizophrenia, Bipolar Disorder, and Schizoaffective disorder." *American Journal of Psychiatry* **164**: 1900-1906.

Marwaha, S. and S. Johnson (2004). "Schizophrenia and employment — a review." *Social Psychiatry and Psychiatric Epidemiology* **39**(5): 337-349.

Mas, A. C. (1994). "Notes on the history of psychiatry in Majorca in the Middle Ages." *History of Psychiatry* **5**: 475-481.

Maschner, H. D. G. (1997). The evolution of Northwest Coast Warfare. *Troubled times: violence and warfare in the past*. D. L. Martin and D. W. Frayer. Amsterdam, Netherlands, Gordon and Breach Publishers.

Mascie-Taylor, C. G. N. (1989). "Spouse similarity for IQ and personality and convergence." *Behavior Genetics* **19**(2): 223-227.

Maslowski, J., D. Jansen van Rensburg and N. Mthoko (1998). "A polydiagnostic approach to the differences in the symptoms of schizophrenia in different cultural and ethnic populations." *Acta Psychiatrica Scandinavica* **98**(1): 41-46.

Mathews, C. A. and V. I. Reus (2001). "Assortative mating in the affective disorders: systematic review and meta-analysis." *Comprehensive Psychiatry* **42**(4): 257-262.

Mayr, E. (2001). *What Evolution Is*. New York, Basic Books.

McClenon, J. (2002). *Wondrous Healing: Shamanism, Human Evolution, and the Origin of Religion*. Dekalb, Illinois, Northern Illinois University Press.

McDonald, C. and K. C. Murphy (2003). "The New Genetics of Schizophrenia." *Psychiatric Clinics of North America* **26**: 41-63.

McGlashan, T. H. (2009). "Psychosis as a Disorder of Reduced Cathetic Capacity: Freud's Analysis of the Schreber Case Revisited." *Schizophrenia Bulletin* **35**(3): 476-481.

McGrath, J. J. (2005). "Myths and Plain Truths about schizophrenia epidemiology — the NAPE lecture 2004." *Acta Psychiatrica Scandinavica* **111**: 4-11.

McGrath, J. J. (2006). "Variations in the Incidence of Schizophrenia: Data Versus Dogma." *Schizophrenia Bulletin* **32**(1): 195-197.

McGrath, J. J., J. Hearle, L. Jenner, K. Plant, A. Drummond and J. M. Barkla (1999). "The fertility and fecundity of patients with psychoses." *Acta Psychiatrica Scandinavica* **99**(6): 441-446.

McGrath, J. J., S. Saha, D. Chant and J. Welham (2008). "Schizophrenia: a concise overview of incidence, prevalence, and mortality." *Epidemiologic Reviews* **30**: 67-76.

McGuire, M. and A. Troisi (1998). *Darwinian Psychiatry*. New York, Oxford University Press.

Menon, M., E. Pomarol-Clotet, P. J. McKenna and R. A. McCarthy (2006). "Probabilistic reasoning in schizophrenia: A comparison of the performance of deluded and nondeluded schizophrenic patients and exploration of possible cognitive underpinnings." *Cognitive Neuropsychiatry* **11**(6): 521-536.

Merikangas, K. R. and D. G. Spiker (1982). "Assortative mating among in-patients with primary affective disorder." *Psychological Medicine* **12**(4): 753-764.

Merimee, T. J., M. B. Grant, C. M. Broder and L. L. Cavalli-Sforza (1989). "Insulin-like growth factor secretion by human B-lymphocytes: a comparison of cells from normal and pygmy subjects." *Journal of Clinical Endocrinology & Metabolism* **69**(5): 978-984.

Merker, B. (2000). Synchronous Chorusing and Human Origins. *The Origins of Music*. N. L. Wallin, B. Merker and S. Brown. Cambridge, MA, MIT Press.

Mesholam-Gately, R., A. J. Guiliano, K. P. Goff, S. V. Faraone and L. J. Seidman (2009). "Neurocognition in First-Episode Schizophrenia: A Meta-Analytic Review." *Neuropsychology* **23**(3): 315-336.

Messias, E. L., C. Chuan-Yu and W. W. Eaton (2007). "Epidemiology of Schizophrenia: Review of Findings and Myths." *Psychiatric Clinics of North America* **30**: 323-338.

Metzner, R. (1998). "Hallucinogenic drugs and plants in psychotherapy and shamanism." *Journal of Psychoactive Drugs* **30**(4): 333-341.

Michelot, D. and L. M. Melendez-Howell (2003). "Amanita muscaria: chemistry, biology, toxicology, and ethnomycology." *Mycological Research* **107**(Pt 2): 131-146.

Middleton, W. C. (1931). "The psychopathology of George Fox, the founder of Quakerism." *Psychological Review* **38**(4): 296-316.

Miles, B. (1997). *Paul McCartney: Many Years from Now*. London, Secker & Warburg.

Miller, G. F. (2001). "Aesthetic Fitness: how sexual selection shaped artistic virtuosity as a fitness indicator and aesthetic preference as mate choice criteria." *Bulletin of Psychology and the Arts* **2**: 20-25.

Miller, G. F. and I. R. Tal (2007). "Schizotypy versus openness and intelligence as predictors of creativity." *Schizophrenia Research* **93**(1-3): 317-324.

Miller, L. E. (2002). "The role of group size in predator foraging decisions or Wedge-capped Capuchin Monkeys (Cebus olivaceus)." *Eat or be Eaten: Predator sensitive foraging among primates*. L. E. Miller. Cambridge, UK, Cambridge University Press.

Miller, R. (1998). *Bare-Faced Messiah: The True Story of L. Ron Hubbard*. New York, Holt.

Millham, A. and S. Easton (1998). "Prevalence of auditory hallucinations in nurses in mental health." *Journal of Psychiatric and Mental health Nursing* **5**: 95-99.

Millon, T. (2004). *Masters of the Mind: Exploring the Story of Mental Illness from Ancient Times to the New Millennium*. Hoboken, NJ, John Wiley and Sons.

Minzenberg, M. J., B. A. Ober and S. Vinogradov (2002). "Semantic priming in schizophrenia: a review and synthesis." *Journal of the International Neuropsychological Society* **8**: 699-720.

Mithen, S. (1996). *The Prehistory of the Mind: a search for the origins of art, religion and science*. London, Thames and Hudson.

Mithen, S. (2005). *The Singing Neanderthals: the origins of music, language, mind and body.* London, Weidenfeld & Nicolson.

Mitteldorf, J. "Death by Design." Retrieved December 4, 2009, from http://www.mathforum.org/\~josh/bydesign.html.

Mitteldorf, J. (2004). "Aging selected for its own sake." *Evolutionary Ecology Research* **6**: 1-17.

Mitteldorf, J. (2006). "Demographic homeostasis and the evolution of senescence." *Evolutionary Ecology Research* **8**: 561-574.

Mitteldorf, J. and J. W. Pepper (2007). "How can evolutionary theory accommodate recent empirical results on organismal senescence?" *Theory In Biosciences* **126**(1): 3-8.

Miyamoto, S., A. S. LaMantia, G. E. Duncan, P. Sullivan, J. H. Gilmore and J. A. Lieberman (2003). "Recent Advances in the Neurobiology of Schizophrenia." *Molecular Interventions* **3**(1): 27-39.

Mohr, S., L. Borras, C. Betrisey, B. Pieree-Yves, C. Gillieron and P. Huguelet (2010). "Delusions with religious content in patients with psychosis: how they interact with spiritual coping." *Psychiatry* **73**(2): 158-172.

Mojtabai, R. and R. A. Nicholson (1995). "Interrater Reliability of Ratings of Delusions and Bizarre Delusions." *American Journal of Psychiatry* **152**(12): 1804-1806.

Moore, M. T., D. Nathan, A. R. Elliot and C. Laubach (1935). "Encephalographic Studies in Mental Disease." *American Journal of Psychiatry* **92**: 43-67.

Moore, O. K. (1957). "Divination: A New perspective." *American Anthropologist* **59**(1): 69-74.

Moritz, S., B. Andresen, C. Perro, M. Schickel, M. Krausz, D. Naber and P. S. Group. (2002). "Neurocognitive performance in first-episode and chronic schizophrenic patients." *European Archives of Psychiatry and Clinical Neuroscience* **252**(1): 33-37.

Moritz, S. and T. S. Woodward (2005). "Jumping to conclusions in delusional and non-delusional schizophrenic patients." *British Journal of Clinical Psychology* **44**(193-207).

Moritz, S. and T. S. Woodward (2006). "A generalized bias against disconfirmatory evidence in schizophrenia." *Psychiatry Research* **142**(2-3): 157-165.

Morrison, A. P. (2002). Cognitive therapy for drug-resistant auditory hallucinations: A case example. *A Casebook of Cognitive Therapy for Psychosis.* A. P. Morrison. New York, Brunner-Routledge.

Mousseau, T. A., B. Sinervo and J. Endler, Eds. (2000). *Adaptive Genetic Variation in the Wild.* New York, Oxford University press.

Mowry, B. J., D. P. Lennon and C. N. De Felice (1994). "Diagnosis of schizophrenia in a matched sample of Australian aborigines." *Acta Psychiatrica Scandinavica* **90**(5): 337-341.

Murdock, G. P. (1934). The Aranda of Central Australia. *Our Primitive Contemporaries.* New York, Macmillan Company: 20-48.

Murphy, H. B. M. and A. C. Raman (1971). "The Chronicity of Schizophrenia in Indigenous Tropical Peoples: Results of a Twelve-Year Follow-up Survey in Mauritius." *British Journal of Psychiatry* **118**: 489-497.

Murphy, J. M. (1976). "Psychiatric Labeling in Cross-Cultural Perspective." *Science* **191**: 1019-1028.

Murray-Jobsis, J. (1991). "An exploratory study of hypnotic capacity of schizophrenic and borderline patients in a clinical setting." *The American Journal of Clinical Hypnosis* **33**(3): 150-160.

Myles-Worsley, M., H. Coon, J. Tiobech, J. Collier, P. Dale, P. Wender, F. Reimherr, A. Polloi and W. Byerley (1999). "Genetic Epidemiological Study of Schizophrenia in Palau, Micronesia: Prevalence and Famiality." *American Journal of Medical Genetics* **88**: 4-10.

Nadkarni, S., V. Arnedo and O. Devinsky (2007). "Psychosis in epilepsy patients." *Epilepsia* **48**(Suppl 9): 17-19.

Nanko, S. and J. Moridaira (1993). "Reproductive rates in schizophrenic outpatients." *Acta Psychiatrica Scandinavica* **87**: 400-404.

Narby, J. and F. Huxley (2001). *Shamans Through Time*. New York, Jeremy P. Tarcher/Putnam.

Nesse, R. M. (2004). "Cliff-edged fitness functions and the persistence of schizophrenia." *Behavioral and Brain Sciences* **27**: 831-885.

Nesse, R. M. (2005). "Maladaptation and Natural Selection." *The Quarterly Review of Biology* **80**(1): 62-70.

Nesse, R. M. and K. C. Berridge (1997). "Psychoactive Drug Use in Evolutionary Perspective." *Science* **278**(5335): 63-66.

Nesse, R. M. and G. C. Williams (1994). *Why We Get Sick*. New York, Vintage Books.

Nestor, P. G., S. J. Akdag, B. F. O'Donnell, M. Niznikiewicz, S. Law, M. E. Shenton and R. W. McCarley (1998). "Word Recall in Schizophrenia: A Connectionist Model." *American Journal of Psychiatry* **155**: 1685-1690.

Nestor, P. G., O. Valdman, M. Niznikiewicz, K. Spencer, R. W. McCarley and M. E. Shenton (2006). "Word priming in schizophrenia: Associational and semantic influences." *Schizophrenia Research* **82**: 139-142.

Nesvag, R., A. Frigessi, E. G. Jonsson and I. Agartz (2007). "Effects of alcohol consumption and antipsychotic medication on brain morphology in schizophrenia." *Schizophrenia Research* **90**(1-3): 52-61.

Nettle, D. (2006a). "Schizotypy and mental health amongst poets, visual artists, and mathematicians." *Journal of Research in personality* **40**: 876-890.

Nettle, D. (2006b). "The evolution of personality variation in humans and other animals." *American Psychologist* **61**(6): 622-631.

Nettle, D. and H. Clegg (2006). "Schizotypy, creativity and mating success in humans." *Proceedings. Biological sciences / The Royal Society* **273**(1586): 611-615.

Newberg, A., E. D'Aquili and V. Rause (2001). *Why God Won't Go Away: Brain Science and the Biology of Belief*. New York, Ballantine Books.

Nicole, L., A. Lesage and P. Lalonde (1992). "Lower incidence and increased male:female ratio in schizophrenia." *British Journal of Psychiatry* **161**: 556-557.

Noll, R. (1989). "What has really been learned about shamanism?" *Journal of Psychoactive Drugs* **21**(1): 47-50.

Nordgaard, J., S. M. Arnfred, P. Handest and J. Parnas (2008). "The Diagnostic Status of First-Rank Symptoms." *Schizophrenia Bulletin* **34**(1): 137-154.
Norris, K. S., B. Wursig, R. S. Wells and M. Wursig (1994). *The Hawaiian Spinner Dolphin*. Berkley, University of California Press.
Nowak, M. A. (2011). *Super Cooperators: Altruism, Evolution, and Why We Need Each Other To Succeed*. New York, Free Press.
O'Carroll, R. E., G. Masterton, N. Dougall, K. P. Ebmeir and G. M. Goodwin (1995). "The Neuropsychiatric Sequelae of Mercury Poisoning: The Mad Hatter's Disease Revisited." *British Journal of Psychiatry* **167**: 95-98.
O'Reilly, T., R. Dunbar and R. Bentall (2001). "Schizotypy and Creativity: an evolutionary connection?" *Personality and Individual Differences* **31**: 1067-1078.
O'Riain, M. J., J. U. M. Jarvis, R. Alexander, R. Buffenstein and C. Peeters (2000). "Morphological castes in a vertebrate." *Proceedings of the National Academy of Sciences* **97**(24): 13194-13197.
Oesterreich, T. K. (1966). *Possesion Demonical and Other Among Primitive Races. Antiquity, The Middle Ages, and Modern Times*. New Hyde Park, New York, University Books.
Ogata, A. and T. Miyakawa (1998). "Religious experiences in epileptic patients with a focus on ictus-related episodes." *Psychiatry and Clinical Neurosciences* **52**: 321-325.
Oldroyd, B. P., H. A. Sylvester, S. Wongsiri and T. E. Rinderer (1994). "Task specialization in a wild bee, *Apis florae* (Hymenoptera: Apidae), revealed by RFLP banding." *Behavioral Ecology and Sociobiology* **34**: 25-30.
Olfson, M., R. Lewis-Fernandez, M. M. Weisman, A. Feder, M. J. Gameroff, D. Pilowsky and M. Fuentes (2002). "Psychotic Symptoms in an Urban General Medicine Practice." *American Journal of Psychiatry* **159**: 1412-1419.
Otsuka, K., A. Sakai and T. Dening (2004). "Haizmann's Madness: the concept of bizarreness and the diagnosis of schizophrenia." *History of Psychiatry* **15**(1): 73-82.
Otterbein, K. F. (2004). *How War Began*. Texas, Texas A&M University Press.
Owen, G. S., G. Cutting and A. S. David (2007). "Are people with schizophrenia more logical than healthy volunteers?" *British Journal of Psychiatry* **191**: 453-454.
Owen, M. J., H. J. Williams and M. C. O'Donovan (2009). "Schizophrenia genetics: advancing on two fronts." *Current Opinion in Genetics & Development* **19**: 266-270.
Page Jr, R. E. and G. E. Robinson (1991). "The Genetics of Division of labour in Honey Bee Colonies " *Advances in Insect Physiology* **23**: 117-169.
Palha, A. P., M. F. Esteves (1997). "The origin of dementia praecox." *Schizophrenia Research* **28**(2-3): 99-103.
Palmer, B. A., V. S. Pankratz and J. M. Bostwick (2005). "The Lifetime Risk of Suicide in Schizophrenia: a reexamination." *Archives of General Psychiatry* **62**: 247-253.
Paloutzian, R. F. and C. L. Park, Eds. (2005). *Handbook of the Psychology of religion and Spirituality*. New York, Guilford Press.

Pantelis, C., D. Velakoulis, S. J. Wood, M. Yucel, A. R. Yung, L. J. Phillips, D. Sun and P. D. McGorry (2007). "Neuroimaging and emerging psychotic disorders: The Melbourne ultra-high risk studies." *International Review of Psychiatry* **19**(4): 373-381.

Pantelis, C., M. Yucel, S. J. Wood, D. Velakoulis, D. Sun, G. Berger, G. W. Stuart, A. Yung, L. Phillips and P. D. McGorry (2005). "Structural Brain imaging Evidence for Multiple Pathological Processes at Different Stages of Brain Development in Schizophrenia." *Schizophrenia Bulletin* **31**(3): 672-696.

Park, G. K. (1963). "Divination and its Social Contexts." *The Journal of the Royal Anthropological Institute of Great Britain and Ireland* **93**(2): 195-209.

Parnas, J. (1988). "Assortative mating in schizophrenia: Results from the Copenhagen high-risk study." *Psychiatry* **51**: 58-64.

Patel, V, T. Musara, T. Butau, P. Maramba and S. Fuyane (1995). "Concepts of mental illness and medical pluralism in Harare." *Psychological Medicine* **25**(3): 485-93.

Patterson, R. and A. Lee (1992). "Louis Riel and the insanity plea that never came." *Canadian Medical Association Journal* **147**(1): 84-87.

Paul, O. (1987). "Da Costa's syndrome or neurocirculatory asthenia." *British Heart Journal* **58**: 306-315.

Pavlov, I. P. (1955). *Pavlov: Selected Works*. Moscow, Foreign Langages Publishing House.

Pearlson, G. D. and B. S. Folley (2008). "Schizophrenia, Psychiatric Genetics and Darwinian Psychiatry: An Evolutionary Framework." *Schizophrenia Bulletin* **34**(4): 722-733.

Pearson, J. L. (2002). *Shamanism and the Ancient Mind: A Cognitive Approach to Archaeology*. Walnut Creek, California, AltaMira Press.

Peers, L. L. (1994). *The Ojibwa of Western Canada, 1780 to 1870* St. Paul, Minnesota Historical Society Press.

Pepperberg, I. M. (1999). *The Alex Studies: Cognitive and Communicative Abilities of Grey Parrots*. Cambridge, MA, Harvard University Press.

Peralta, V. and M. J. Cuesta (1999). "Diagnostic significance of Schneider's first-rank symptoms in schizophrenia. Comparative study between schizophrenic and non- schizophrenic psychotic disorders." *The British Journal of Psychiatry* **174**: 243-248.

Peralta, V. and M. J. Cuesta (2005). "The Underlying Structure of Diagnostic Systems of Schizophrenia: a comprehensive polydiagnostic approach." *Schizophrenia Research* **79**(2-3): 217-229.

Peralta, V. and M. J. Cuesta (2007). "A dimensional and categorical architecture for the classification of psychotic disorders." *World Psychiatry* **6**(2): 100-101.

Perkins, R. and M. Rinaldi (2002). "Unemployment rates among patients with long-term mental health problems: A decade of rising unemployment." *Psychiatric Bulletin* **26**: 295-298.

Perr, I. N. (1992). "Religion, Political Leadership, Charisma, and Mental Illness: The Strange Story of Louis Riel." *Journal of Forensic Sciences* **37**(2): 574-584.

Peters, E. R., S. A. Joseph and P. A. Garety (1999). "Measurement of Delusional Ideation in the Normal Population: Introducing the PDI (Peters et al. Delusions Inventory)." *Schizophrenia Bulletin* **25**(3): 553-576.

Peters, L. G. and D. Price-Williams (1980). "Towards an experiential analysis of shamanism." *American Ethnologist* **7**(3): 397-418.

Pettinati, H. M., L. G. Kogan, F. J. Evans, J. H. Wade, R. L. Horne and J. M. Staats (1990). "Hypnotizability of psychiatric inpatients according to two different scales." *American Journal of Psychiatry* **147**(1): 69-75.

Pfau, A. N. (2008). *Madness in the Realm: Narratives of mental Illness in late Medieval France.* PhD University of Michigan, Horace H. Rackham School of Graduate Studies.

Pfeifer, S. (1994). "Belief in demons and exorcism in psychiatric patients in Switzerland." *British Journal of Medical Psychology* **67**(3): 247-258.

Pfeifer, S. (1999). "Demonic attributions in nondelusional disorders." *Psychopathology* **32**(5): 252-259.

Philo, C. (2003). *A geographical history of institutional provision for the insane from medieval times to the 1860s in England and Wales: The space reserved for insanity.* New York, Edwin Mellen Press.

Pickover, C. A. (2009). *The Loom of God: Tapestries of Mathematics and Mysticism.* New York, Sterling Publishing.

Pierre, J. M. (2010). "Hallucinations in Nonpsychotic Disorders: Toward a Differential Diagnosis of 'Hearing Voices'." *Harvard Review of Psychiatry* **22**-35.

Pierri, J. N., A. S. Chaudry, T. W. Woo and D. A. Lewis (1999). "Alterations in Chandelier Neuron Axon Terminals in the Prefrontal Cortex of Schizophrenic Subjects." *American Journal of Psychiatry* **156**: 1709-1719.

Place, E. J. and G. C. Gilmore (1980). "Perceptual Organization in Schizophrenia." *Journal of Abnormal Psychology* **89**(3): 409-418.

Plotkin, M. J. (1993). *Tales of a Shaman's Apprentice.* New York, Viking.

Pluess, M. and J. Belsky (2009). "Differential susceptibility to rearing experience: the case of childcare." *Journal of Child Psychology and Psychiatry* **50**(4): 396-404.

Polimeni, J., D. W. Campbell, D. Gill, B. Sawatsky and J. P. Reiss (2010). "Diminished Humor Perception in Schizophrenia: relationship to social and cognitive functioning." *Journal of Psychiatric Research* **44**(7): 434-440.

Polimeni, J., J. Reiss and J. Sareen (2005). "Could obsessive-compulsive disorder have originated as a group selected adaptive trait in traditional societies?" *Medical Hypotheses* **65**: 655-664.

Polimeni, J. and J. P. Reiss (2002). "How Shamanism and Group Selection may reveal the origins of schizophrenia." *Medical Hypotheses* **58**: 244-248.

Polimeni, J. and J. P. Reiss (2003). "Evolutionary Perspectives on Schizophrenia." *Canadian Journal of Psychiatry.* **48**(1): 34-39.

Polimeni, J. and J. P. Reiss (2006). "Humor Perception Deficits in Schizophrenia." *Psychiatry Research* **141**(2): 229-232.

Polimeni, J. and J. P. Reiss (2006). "The First Joke: Exploring the evolutionary origins of humor." *Evolutionary Psychology* **4**: 347-366.

Popper, K. R. (1959). *The Logic of Scientific Discovery.* New York, Basic Books.

Porter, R. (2002). *Madness: A Brief History.* Oxford, Oxford University Press.

Posey, T. B. and M. E. Losch (1984). "Auditory hallucinations of hearing voices in 375 normal subjects " *Imagination, Cognition and Personality* **3**(2): 99-113.

Post, F. (1994). "Creativity and Psychopathology: A study of 291 World-Famous Men." *British Journal of Psychiatry* **165**: 22-34.

Prentice, K. J., J. M. Gold and W. T. J. Carpenter (2005). "Optimistic bias in the perception of personal risk: patterns in schizophrenia." *American Journal of Psychiatry* **162**(3): 507-512.

Preti, A. and P. Miotto (1997). "Creativity, Evolution and Mental Illnesses." *Journal of Memetics –Evolutionary Models of Information Transmission* **1**: 97-104.

Price, J.S. (2010). The culture of religious belief systems and changes of belief system. *Politics and Culture*. April 29. Part of a symposium edited by Joseph Carroll.

Price, J. (2009). The adaptiveness of changing religious belief systems. *The Biology of Religious Behaviour: The evolutionary origins of faith and religion*. J. R. Feierman. Santa Barbara, CA, Praeger.

Price, J., R. Gardner and M. Erickson (2004). "Can depression, anxiety and somatisation be understood as appeasement displays?" *Journal of Affective Disorders* **79**: 1-11.

Puurtinen, M. and T. Mappes (2009). "Between-Group competition and human cooperation." *Proceedings of the Royal Society - Biological Sciences* **276**(1655): 355-360.

Radin, P. (1937). *Primitive religion: Its nature and origin*. New York, Dover.

Rank, G. (1967). Shamanism as a Research Subject: Some Methodological Viewpoints. *Studies in Shamanism*. C. M. Edsman. Stockholm, Almqvist: 15-22.

Regier, D. A., M. E. Farmer, D. S. Rae, B. Z. Locke, S. J. Keith, L. L. Judd and F. K. Goodwin (1990). "Comorbidity of Mental Disorders With Alcohol and Other Drug Abuse: Results From the Epidemiological Catchment Are (ECA) Study." *JAMA* **264**(19): 2511-2518.

Reichel-Dolmatoff, G. (1972). The Cultural context of an Aboriginal Hallucinogen: Banisteropsis Caapi. *Flech of the Gods: The Ritual use of Hallucinogens*. P. T. Furst. Long Grove, Illinois, Waveland Press.

Reilly, T. and J. Scott (1993). "Effects of elevating blood alcohol levels on tasks related to dart throwing." *Perceptual & Motor Skills* **77**(1): 25-26.

Reiss, J. P., J. Polimeni, D. W. Campbell, J. Sareen, M. P. Paulus, L. N. Ryner and R. B. Bolster (2007). Word Association in Schizophrenia: An fMRI Study. *Canadian Psychiatric Association Annual Meeting*. Montreal, Canada.

Renvoize, E. B. and A. W. Beveridge (1989). "Mental illness and the late Victorians: a study of patients admitted to three asylums in York, 1880-1884." *Psychological Medicine* **19**(1): 19-28.

Reynolds, C. A., L. A. Baker and N. L. Pedersen (1996). "Models of spouse similarity: Applications to fluid ability measured in twins and their spouses." *Behavior Genetics* **26**(2): 73-88.

Richards, R., D. K. Kinney, I. Lunde, M. Benet and A. P. C. Merzel (1988). "Creativity in Manic-Depressives, Cyclothymes, Their Normal Relatives, and Control Subjects." *Journal of Abnormal Psychology* **97**(3): 281-288.

Richards, R. L. (1981). "Relationships between creativity and psychopathology: an evaluation and interpretation of the evidence." *Genetic Psychology Monographs* **103**(second half): 261-324.

Richey, S. W. (2000). "Joan of Arc: A Military Appreciation." from http://www.stjoan-center.com/military/stephenr.html.
Ridley, M. (1996). *Evolution*. Cambridge, MA, Blackwell Science.
Ridley, M. (2001). *The Cooperative Gene: How Mendel's Demon Explains the Evolution of Complex Beings*. New York, The Free Press.
Riedweg, C. (2005). *Pythagoras: His Life, Teaching, and Influence*. Ithaca, New York, Cornell University Press.
Riley, B. and K. S. Kendler (2006). "Molecular genetic studies of schizophrenia." *European Journal of Huamn Genetics* **14**(6): 669-680.
Riley, I. W. (1903). *The Founder of Mormonism: A Psychological Study of Joseph Smith, Jr*. London, William Heinemann.
Ringen, P. A., A. Vaskinn, K. Sundet, J. A. Engh, H. Jonsdottir, C. Simonsen, S. Friis, S. Opjordsmoen, I. Melle and O. A. Andreassen (2010). "Opposite relationships between cannabis use and neurocognitive functioning in bipolar disorder and schizophrenia." *Psychological Medicine* **40**(8): 1337-1347.
Rios-Cardenas, O., M. S. Tudor and M. R. Morris (2007). "Female preference variation has implications for the maintenance of an alternative mating strategy in a swordtail fish." *Animal Behaviour* **74**: 633-640.
Ritchie, M. A. (1996). *Spirit of the Rainforest: A Yanomamo Shaman's story*. Chicago, Island Lake Press.
Robinson, B. W. and D. Schluter (2000). Natural Selection and the Evolution of Adaptive Genetic Variation in Norther Freshwater Fishes. *Adaptive Genetic Variation in the Wild*. T. A. Mousseau, B. Sinervo and J. Endler. New York, Oxford University Press.
Robinson, G. E. (1992). "Regulation of Division of Labor in Insect Societies." *Annual Review of Entomology* **37**: 637-665.
Rodman, D. M. and S. Zamudio (1991). "The cystic fibrosis heterozygote — advantage in surviving cholera?" *Medical Hypotheses* **36**(3): 253-258.
Rodriguez-Sanchez, J. M., R. Ayesa-Arriola, I. Mata, T. Moreno-Calle, R. Perez-Iglesias, C. Gonzalez-Blanch, J. A. Perianez, J. L. Vazquez-Barquero and B. Crespo-Facorro (2010). "Cannabis use and cognitive functioning in first-episode schizophrenia patients." *Schizophrenia Research*.
Rogers, S. A. and R. F. Paloutzian (2006). Schizophrenia, Neurology, and Religion: What Can Psychosis Teach Us about the Evolutionary Role of Schizophrenia? *Where God and Science Meet: How Brain and Evolutionary Studies Alter Our Understanding of Religion*. P. McNamara. Westport, Conn, Praeger. **3**.
Rolling Stone. (2006). from http://www.rollingstone.com/music/news/syd-barrett-1946-2006-20060711
Ropacki, S. A. and D. V. Jeste (2005). "Epidemiology of and risk factors for psychosis of Alzheimer's disease: a review of 55 studies published from 1990 to 2003." *American Journal of Psychiatry* **162**(11): 2022-2030.
Rosenberg, N. (2010). Brooklyn Actor Held in Killing of his Mother. *New York Times*. New York.
Rosenthal, D. (1975). "Discussion: The concept of subschizophrenic disorders." *Proceedings of the annual meeting of the American Psychopathological Association* **63**: 200-208.

Roser, P., F. X. Vollenweider and W. Kawohl (2010). "Potential antipsychotic properties of central cannabinoid (CB1) receptor antagonists." *The World Journal of Biological Psychiatry* **11**(2): 208-219.

Rossano, M. J. (2010). *Supernatural selection: How religion evolved.* Oxford, Oxford University Press.

Roy, A. K., S. V. Rajesh, N. Iby, J. M. Jose and G. R. Sarma (2003). "A study of epilepsy-related psychosis." *Neurology India* **51**(3): 359-360.

Rudaleviciene, P., T. Stompe, A. Narbekovas, N. Raskauskiene and R. Bunevicius (2008). "Are religious delusions related to religiosity in schizophrenia?" *Medicina (Kaunas)* **44**(7): 529-535.

Rushton, J. P. (1990). "Creativity, Intelligence, and Psychoticism." *Personality and Individual Differences* **11**(12): 1291-1298.

Russell, A. J., J. C. Munro, P. B. Jones, D. R. Hemsley and R. M. Murray (1997). "Schizophrenia and the Myth of Intellectual Decline." *American Journal of Psychiatry* **155**(3): 325-336.

Ryan, R. E. (1999). *The Strong Eye of the Shaman: A journey into the caves of consciousness.* Rochester, Vermont, Inner Traditions.

Sacco, R. L., M. Elkind, B. Boden-Albala, I. F. Lin, D. E. Kargman, A. Hauser, S. Shea and M. C. Paik (1999). "The Protective Effect of Moderate Alcohol Consumption on Ischemic Stroke." *JAMA* **281**(1): 53-60.

Sachdev, P. (1998). "Schizophrenia-like psychosis and epilepsy: The status of the association." *American Journal of Psychiatry* **155**(3): 325-336.

Sachdev, P. S. and M. S. Keshaven, Eds. (2010). *Secondary Schizophrenia.* Cambridge, Cambridge University Press.

Sagud, M., A. Milhaijevic-Peles, D. Muck-Seler, N. Pivac, B. Vuksan-Cusa, T. Bratal-jenovic and M. Jakovljevic (2009). "Smoking and schizophrenia." *Psychiatria Danubina* **21**(3): 371-375.

Saha, S., D. Chant, J. Welham and J. McGrath (2005). "A Systematic Review of Prevalence of Schizophrenia." *PLoS Medicine* **2**(5): e141.

Sasaki, Y. (1969). Psychiatric Study of the Shaman in Japan. *Mental Health Research in Asia and the Pacific.* W. Caudill and T. Y. Lin. Honolulu, East-West Center Press.

Saumier, D. and H. Chertkow (2002). "Semantic Memory " *Current Neurology and Neuroscience Reports* **2**(6): 516-522.

Schatzberg, A. F., J. O. Cole and C. DeBattista (2005). *Manual of Clinical Psychopharmacology.* Washington, DC, American Psychiatric Publishing.

Schmid-Burgk, W., W. Becker, V. Diekmann, R. Jurgens and H. H. Kornhuber (1982). "Disturbed smooth pursuit and saccadic eye movements in schizophrenia." *Archiv für Psychiatrie und Nervenkrankheiten* **232**: 381-389.

Schmitt, A., C. Steyskal, H. G. Berstein, T. Schnieder-Axmann, E. Parlapani, E. L. Schaeffer, W. F. Gattaz, B. Bogerts, C. Schmitz and P. Falkai (2009). "Stereologic investigation of the posterior part of the hippocampus in schizophrenia." *Acta Neuropathologica* **117**: 395-407.

Schneider, K. (1959). *Clinical Psychopathology.* New York, Grune & Stratton.

Schreiber, S. (2010). "Demons and spirits in the 3rd millennium — the haunting goes on." *WevmedCentral PSYCHIATRY* **1**(9): WMC00673.

Schroeder, S. A., D. M. Gaughan and M. Swift (1995). "Protection against bronchial asthma by CFTR F508 mutation: A heterozygote advantage in cystic fibrosis." *Nature Medicine* **1**(7): 703-705.

Schuldberg, D. (1997). Scizotypal and Hypomanic Traits, Creativity, and Psychological Health. *Eminent Creativity, Everyday Creativity and Health*. M. A. Runco and R. Richards, Ablex Publishing.

Schuldberg, D. (2000). "Six Subclinical Spectrum Traits in Normal Creativity." *Creativity Research Journal* **13**(1): 5-16.

Schuldberg, D., C. French, L. Stone and J. Heberle (1988). "Creativity and Schizotypal Traits: Creativity Test Scores and Perceptual Aberration, Magical Ideation, and Impulsive Nonconformity." *Journal of Nervous and Mental Disease* **176**(11): 648-657.

Schultes, R. E. (1972). An Overview of Hallucinogens in the Western Hemisphere. *Flesh of the Gods: The Ritual use of Hallucinogens*. P. T. Furst. Long Grove, Illinois, Waveland Press.

Seeley, T. D. (1997). "Honey Bee Colonies are group-level adaptive units." *The American Naturalist* **150**(Suppl Multilevel Selection): s22-s41.

Selzer, J. A. and J. A. Lieberman (1993). "Schizophrenia and Substance Abuse." *Psychiatric Clinics of North America* **16**(2): 401-412.

Sharma, T. and P. Harvey (2000). *Cognition in Schizophrenia: Impairments, Importance and Treatment Strategies*. Oxford, Oxford University Press.

Shea, B. T. and R. C. Bailey (1996). "Allometry and adaptation of body proportions and stature in African pygmies." *American Journal of Physical Anthropology* **100**(3): 311-340.

Shirokogoroff, S. M. (1935). *Psychomental Complex of the Tungus* London, Kegan Paul, Trench, Trubner & Co.

Shorter, E. (1997). *A History of Psychiatry: From the era of the asylum to the age of Prozac*. New York, John Wiley and Sons.

Shorter, E. (2005). *A Historical Dictionary of Psychiatry*. Oxford, Oxford University Press.

Sickinger, R. L. (2000). "Hitler and the Occult: The Magical Thinking of Adolf Hitler." *The Journal of Popular Culture* **34**(2): 107-125.

Siddle, R., G. Haddock, N. Tarrier and E. B. Faragher (2002). "Religious delusions in patients admitted to hospital with schizophrenia." *Social Psychiatry and Psychiatric Epidemiology* **37**(3): 130-138.

Sieroszewski, W. (1993). The Yakut: an experiment in ethnographic research. *'Rossiiskaia polit. entsiklopediia'*. Moskva.

Silventoinen, K., J. Kaprio, E. Lahelma, R. J. Viken and R. J. Rose (2003). "Assortative mating by body height and BMI: Finnish twins and their spouses." *American Journal of Human Biology* **15**(5): 620-627.

Silverman, J. (1967). "Shamans and Acute Schizophrenia." *American Anthropologist* **69**(1): 21-31.

Simeonova, D. I., K. D. Change, C. Strong and T. A. Ketter (2005). "Creativity in familial bipolar disorder " *Journal of Psychiatric Research* **39**: 623-631.

Smith, C. U. M. (1996). *Elements of Molecular Neurobiology* Chichester, UK John Wiley & Sons.

Smith, D. B. (2007). *Muses, Madmen, and Prophets: Rethinking the history, science, and meaning of auditory hallucination.* New York, Penguin Press.

Smolker, R. (2001). *To Touch a Wild Dolphin: a journey of discovery with the sea's most intelligent creatures.* New York, Nan A. Talese Doulbleday.

Smoller, J. W. and C. T. Finn (2003). "Family, Twin and Adoption Studies of bipolar disorder." *American journal of medical genetics. Part C, Seminars in medical genetics* **123C**(1): 48-58.

Sober, E. and D. S. Wilson (1998). *Unto Others: The evolution and psychology of unselfish behavior.* Cambridge, MA, Harvard University Press.

Somasundaram, D., T. Thivakaran and D. Bhugra (2008). "Possession states in Northern Sri Lanka." *Psychopathology* **41**(4): 245-253.

Somnath, C. P., R. Janardhan, Y.C. and S. Jain (2002). "Is there familial overlap between schizophrenia and bipolar disorder?" *Journal of Affective Disorders* **72**(3): 243-247.

Sosis, R. (2004). "The Adaptive Value of Religious Ritual." *American scientist* **92**(2): 166-172.

Sosis, R. and C. Alcorta (2003). "Signaling, Solidarity, and the Sacred: The Evolution of Religious Behavior." *Evolutionary Anthropology* **12**: 264-274.

Sosis, R., H. C. Kress and J. S. Boster (2007). "Scars for war: evaluating alternative signaling explanations for cross-cultural variance in ritual costs." *Evolution and Human Behavior* **28**: 234-247.

Spilka, B., R. W. J. Hood, B. Hunsberger and R. Gorsuch, Eds. (2003). *The Psychology of Religion.* New York, Guilford Press.

Spitzer, M. (1992). Word-Associations in Experimental Psychiatry: A Historical perspective. *Phenomenology, Language and Schizophrenia.* M. Spitzer, F. Uehlein, M. A. Schwartz and C. Mundt, Springer-Verlag: 160-196.

Spitzer, M. (1997). "A Cognitive Neuroscience View of Schizophrenic Thought Disorder." *Schizophrenia Bulletin* **23**(1): 29-50.

Spyropoulos, B. (1988). "Tay-Sachs carriers and Tuberculosis resistance." *Nature* **331**(6158): 666.

Srinivasan, T. N. and R. Padmavati (1997). "Fertility and schizophrenia: evidence for increased fertility in the relatives of schizophrenic patients." *Acta Psychiatrica Scandinavica* **96**(4): 260-264.

St. Clair, D. (2009). "Copy Number Variation and Schizophrenia." *Schizophrenia Bulletin.* **35**(1): 9-12.

Stahl, P. W. (1986). "Hallucinatory Imagery and the Origin of Early South American Figurine Art." *World Archaeology* **18**(1): 134-150.

Stander, P. E. (1992). "Cooperative hunting in lions: the role of the individual." *Behavioral Ecology and Sociobiology* **29**: 445-454.

Stark, R. (1999). "A Theory of Revelations." *Journal for the Scientific Study of Religion* **38**(2): 287-308.

Steadman, L. B. and C. T. Palmer (2008). *The Supernatural and Natural Selection: The Evolution of Religion.* Boulder, Colorado Paradigm Publishers.

Stearns, P. N. (1994). *Encyclopedia of Social History.* Garland Publishing 780.

Steen, R. G., R. M. Hamer and J. A. Lieberman (2005). "Measurement of brain metabolites by 1H magnetic resonance spectroscopy in patients with schizophrenia: a systematic review and meta-analysis." *Neuropsychopharmacology* **30**: 1949-1962.

Steinberg, S. and J. Weiss (1954). "The Art of Edvard Munch and its Function in his Mental Life." *Psychoanalytic Quarterly* **23**: 409-423.

Stephen, M. and L. K. Suryani (2000). "Shamanism, psychosis and autonomous imagination." *Culture, Medicine and Psychiatry* **24**(1): 5-40.

Stevens, A. and J. Price (1996). *Evolutionary Psychiatry: a new beginning*. London Routledge.

Stevens, A. and J. Price (2000). *Prophets, Cults and Madness*. London, Duckworth-Publishers.

Stip, E. and G. Letourneau (2009). "Psychotic symptoms as a continuum between normality and pathology." *Canadian Journal of Psychiatry* **54**(3): 140-151.

Stompe, T., A. Friedman, G. Ortwein, R. Strobl, H. R. Chaudhry, N. Najam and M. R. Chaudhry (1999). "Comparison of Delusions among schizophrenic in Austria and in Pakistan." *Psychopathology* **32**(5): 225-234.

Stone, M. H. (1997). *Healing the Mind: A History of Psychiatry from Antiquity to the Present*. New York, W.W. Norton & Co.

Storr, A. (1996). *Feet of Clay: A study of Gurus*. London, HarperCollins.

Striedter, G. F. (2005). *Principles of Brain Evolution*. Sunderland, MA, Sinaur Associates Inc Publishers.

Stromgren, E. (1987). "Changes in the incidence of schizophrenia." *British Journal of Psychiatry* **150**: 1-7.

Sullivan, P. F. (2005). "The Genetics of Schizophrenia." *PLoS Medicine* **2**(7): e212.

Sullivan, R. J. and E. H. Hagen (2002). "Psychotropic substance-seeking: evolutionary pathology or adaptation?" *Addiction* **97**: 389-400.

Sutrala, S. R., D. Goossens, N. M. Williams, L. Heyrman, R. Adolfsson, N. Norton, P. R. Buckland and J. Del-Favero (2007). "Gene Copy Number Variation in Schizophrenia." *Schizophrenia Research* **96**: 93-99.

Suvisaari, J. M., J. K. Haukka, A. J. Tanskanen and J. K. Lonnqvist (1999). "Decline in theincidence of schizophrenia in Finnish cohorts born from 1954-1965." *Archives of General Psychiatry* **56**: 733-740.

Svensson, A. C., P. Lichtenstein, S. Sandin and C. M. Hultman (2007). "Fertility of first-degree relative of patients with schizophrenia: A three generation perspective." *Schizophrenia Research* **91**(1-3): 238-245.

Szasz, T. (1970). *The Manufacture of Madness: A comparative study of the Inquisition and the Mental Health Movement*. Syracuse, NY, Syracuse University press.

Szasz, T. (1974). *The Myth of Mental Illness: Foundations of a theory of Personal Conduct*. New York, HarperCollins.

Taber, K. H. and R. A. Hurley (2007). "Neuroimaging in Schizophrenia: Misattributions and Religious Delusions." *The Journal of Neuropsychiatry and Clinical Neurosciences* **19**: 1.

Tam, G. W. C., R. Redon, N. P. Carter and S. G. N. Grant (2009). "The Role of DNA Copy Number Variation in Schizophrenia " *Biological Psychiatry* **66**: 1005-1012.

Tamminga, C. A. (1997). "Gender and Schizophrenia." *Journal of Clinical Psychiatry* **58**(Suppl 15): 33-37.

Tamminga, C. A., P. J. Sirovatka, D. A. Regier and J. van Os (2010). *Deconstructing Psychosis: Refining the Research Agenda for DSM-V*. Arlington, Virginia, American Psychiatric Association.

Tanner, A. (1978). Divination and Decisions: multiple explanations for Algonkian scapulimancy. *The Yearbook of Symbolic Anthropology*. E. Schwimmer. Montreal, McGill-Queens University press.

Tateyama, M., M. Asai, M. Hashimoto, M. Bartels and S. Kasper (1998). "Transcultural Study of Schizophrenic Delusions." *Psychopathology* **31**: 59-68.

Tattersall, I. (1995). *The Fossil Trail: How we know what we think we know about human evolution*. Oxford, Oxford University Press.

Taylor, D. C. and M. Lochery (1987). "Temporal lobe epilepsy: origin and significance of simple and complex auras." *Journal of Neurology, Neurosurgery, and Psychiatry* **50**: 673-681.

Tedlock, B. (2005). *The Woman in the Shaman's Body*. New York, Bantam Books.

Teuton, J., R. Bentall and C. Dowrick (2007). "Conceptualizing Psychosis in Uganda: The Perspective of Indigenous and Religious Healers." *Transcultural psychiatry* **44**(1): 79-114.

Thalbourne, M. A. (1994). "Belief in the paranormal and its relationship to schizophrenia-relevant measures: a confirmatory study." *British Journal of Clinical Psychology* **33**: 78-80.

Thomas, K. (1992). *Religion and the Decline of Magic*. London, Penguin Books.

Thurn, E. (1883). *Among the Indians of Guiana*. New York, Dover.

Tien, A. Y. (1991). "Distributions of hallucinations in the population." *Social Psychiatry and Psychiatric Epidemiology* **26**: 287-292.

Tienari, P. and L. C. Wynne (1994). "Adoption studies of schizophrenia." *Annals of Medicine* **26**(4): 233-237.

Tienari, P., L. C. Wynne, J. Moring, I. Lahti, M. Naarala, A. Sorri, K. E. Wahlberg, O. Saarento, M. Seitamaa and M. Kaleva (1994). "The Finnish adoptive family study of schizophrenia. Implications for family research." *British Journal of Psychiatry* **23**: 20-26.

Tienari, P., L. C. Wynne, J. Moring, K. Läksy, P. Nieminen, A. Sorri, I. Lahti, K. E. Wahlberg, M. Naarala, K. Kurki-Suonio, O. Saarento, P. Koistinen, T. Tarvainen, H. Hakko and J. Miettunen (2000). "Finnish adoptive family study: sample selection and adoptee DSM-III-R diagnoses." *Acta Psychiatrica Scandinavica* **101**: 433-443.

Torrey, E. F. (1987). "Prevalence studies in schizophrenia." *British Journal of Psychiatry* **150**: 598-608.

Torrey, E. F. and J. Miller (2001). *The Invisible Plague: the rise of mental illness from 1750 to the present*. New Jersey, Rutgers University Press.

Torrey, E. F., B. B. Torrey and B. G. Burton-Bradley (1974). "The Epidemiology of Schizophrenia in Papua New Guinea." *American Journal of Psychiatry* **131**(5): 567-573.

Tregellas, J. R., S. Shatti, J. L. Tanabe, L. F. Martin, L. Gibson, K. Wylie and D. C. Rojas (2007). "Gray matter volume differences and the effects of smoking on gray matter in schizophrenia." *Schizophrenia Research* **97**(1-3): 242-249.

Tsoi, D. T., K. Lee, K. A. Gee, K. L. Holden, R. W. Parks and P. W. Woodruff (2008). "Humour experience in schizophrenia: relationship with executive dysfunction and psychosocial impairment." *Psychological Medicine* **38**: 801-810.

Turbott, J. (1997). "The Meaning and function of ritual in psychiatric disorder, religion and everyday behavior." *Australian and New Zealand Journal of Psychiatry* **31**: 835-843.

Turner, T. H. (1992). "Schizophrenia as a permanent problem: Some aspects of historical evidence in the recency (new disease) hypothesis." *History of Psychiatry* **3**: 413-429.

Tye, S. J., M. A. Frye and L. K.H. (2009). "Disrupting Disordered Neurocircuitry: Treating Refractory Psychiatric Illness With Neuromodulation." *Mayo Clinic Proceedings* **84**(6): 522-532.

Uhlhaas, P. J. and A. L. Mishara (2007). "Perceptual Anomalies in Schizophrenia: Integrating Phenomenology and Cognitive Neuroscience." *Schizophrenia Bulletin* **33**(1): 142-156.

Uhlhaas, P. J., W. A. Phillips, G. Mitchell and S. M. Silverstein (2006). "Perceptual Grouping in Disorganized Schizophrenia." *Psychiatry Research* **145**(105-117).

Uhlhaas, P. J. and S. M. Silverstein (2005). "Perceptual Organization in Schizophrenia Spectrum Disorders: Empirical Research and Theoretical Implications." *Psychological Bulletin* **131**(4): 618-632.

Umbricht, D., G. Degreef, W. B. Barr, J. A. Lieberman, S. Pollack and N. Schaul (1995). "Postictal and chronic psychoses in patients with temporal lobe epilepsy." *American Journal of Psychiatry* **152**(2): 224-231.

Valentine, V. F. and F. G. Vallee (1968). *Eskimo of the Canadian Artic*. Toronto, McClelland and Stewart.

Vallee, B. (1997). *Alcohol and the Development of Human Civilization. In Exploring the Universe*. Oxford, Oxford University Press.

Van Crevald, M. (1991). *The Transformation of War*. New York, The Free Press.

van der Feltz-Cornelis, C. M., A. P. Aldenkamp, H. J. Ader, A. Boenink, D. Linszen and R. Van Dyck (2008). "Psychosis in epilepsy patients and other chronic medically ill patients and the role of cerebral pathology in the onset of psychosis: a clinical epidemiological study." *Seizure* **17**(5): 446-456.

Van Doren, C. A. (1991). *A History of Knowledge*. New York, Ballantine Books.

van Haren, N. E., H. G. Schnack, W. Cahn, M. P. van den Heuvel, C. Lepage, L. Collins, A. C. Evans, H. E. Hulshoff and R. S. Kahn (2011). "Changes in cortical thickness during the course of illness in schizophrenia." *Archives of General Psychiatry* **68**(9): 871-880.

van Os, J. (2003). "Is there a continuum of psychotic experiences in the general population?" *Epidemiologia e psichiatria sociale* **12**(4): 242-252.

van Os, J. and H. Verdoux (2003). Diagnosis and Classification of Schizophrenia: categories versus dimensions, distributions versus disease. *The Epidemiology of Schizophrenia*. R. M. Murray, P. B. Jones, E. Susser, J. van Os and M. Cannon. Cambridge, UK, Cambridge University Press.

van Os, J., H. Verdoux, S. Maurice-Tison, B. Gay, F. Liraud, R. Salamon and M. Bourgeois (1999). "Self-reported psychosis-like symptoms and the continuum of psychosis." *Social Psychiatry and Psychiatric Epidemiology* **34**(9): 459-463.

Veatch, T. C. (1998). "A theory of humor." *Humor* **11**: 161-215.

Veblen, T. (1899). *The Theory of the Leisure Class*. New York, The Modern Library.

Verdoux, H., S. Maurice-Tison, B. Gay, J. van Os, R. Salamon and M. L. Bourgeois (1998). "A Survey of Delusional ideation in Primary-Care Patients." *Psychological Medicine* **28**: 127-134.

Vitebsky, P. (1995). *The Shaman: Voyages of the Soul Trance, Ecstasy and Healing from Siberia to the Amazon*. London, Duncam Baird Books.

Vlachos, I. O., B. Stavroula and P. Hartocollis (1997). "Magico-religious Beliefs and Psychosis." *Psychopathology* **30**: 93-99.

Vogel, H. P. (1979). "Fertility and Sibship Size in a psychiatric patient population." *Acta Psychiatrica Scandinavica* **60**: 483-503.

Vogiatzoglou, A., H. Refsum, C. Johnston, S. M. Smith, K. M. Bradley, C. de Jager, M. M. Budge and A. D. Smith (2008). "Vitamin B12 status and rate of brain volume loss in community-dwelling elderly." *Neurology* **71**(11): 826-832.

Vollenweider, F. X., M. F. Vollenweider-Scherpenhuyzen, A. Babler, H. Vogel and D. Hell (1998). "Psilocybin induces schizophrenia-like psychosis in humans via a serotonin-2 agonist action." *Neuroreport* **9**(17): 3897-3902.

Vollenweider, F. X., P. Vontobel, D. Hell and K. L. Leenders (1999). "5-HT modulation of dopamine release in basal ganglia in psilocybin-induced psychosis in man — a PET study with (11C) raclopride." *Neuropsychopharmacology* **20**(5): 424-433.

Von Frisch, K. (1953). *The Dancing Bees*. New York, Harcourt, Brace & Co.

Waddell, C. (1998). "Creativity and mental illness: Is there a link?" *Canadian Journal of Psychiatry* **43**: 166-172.

Waddington, J. L. and H. A. Youssef (1994). "Evidence for a gender-specific decline in the rate of schizophrenia in rural Ireland over a 50-year period." *British Journal of Psychiatry* **164**: 171-176.

Waddington, J. L. and H. A. Youssef (1996). "Familial-genetic and reproductive epidemiology of schizophrenia in rural Ireland: Age at onset, familial morbid risk and parental fertility." *Acta Psychiatrica Scandinavica* **93**(1): 62-68.

Wade, M. J. (1976). "Group selection among laboratory populations of *Tribolium*." *Proceedings of the National Academy of Sciences* **73**: 4604-4607.

Wade, M. J. (1977). "An experimental study of group selection." *Evolution* **31**: 134-153.

Wade, M. J. (1978). "A critical review of the models of group seletion." *The Quarterly Review of Biology* **53**: 101-114.

Wade, N. (2006). *Before the Dawn: Recovering the lost history of our ancestors*. New York, Penguin Books.

Wade, N. (2009). *The Faith Instinct: How Religion evolved and why it endures*. New York, Penguin press.

Wagley, C. (1977). *Welcome of Tears: The Tapirapé Indians of Central Brazil*. New York, Oxford University press.

Wakefield, M. (1991). *Understanding Scientology*. Tampa, Florida, Coalition of Concerned Citizens.

Waldo, M. C. (1999). "Schizophrenia in Kosrae, Micronesia: Prevalence, gender ratios, and clinical symptomology." *Schizophrenia Research* **35**(2): 175-181.

Waldron, H. A. (1983). "Did the Mad Hatter have Mercury Poisoning?" *British Medical Journal* **287**(6409): 1961.

Waller, N. G., B. A. Kojetin, T. J. J. Bouchard, D. T. Lykken and A. Tellegen (1990). "Genetic and Environmental Influences on religious Interests, Attitudes, and Values: A study of twins reared apart and together." *Psychological Science* **1**: 138-142.

Walsh, R. (1997). "The Psychological Health of HSamans: A Reevaluation." *Journal of the American Academy of Religion* **65**(1): 101-124.

Walsh, R. (2007). *The World of Shamanism: New Views of an Ancient Tradition.* Woodbury, Minnesota Llewellyn Publications.

Ward, C. and M. H. Beaubrun (1981). "Spirit possession and neuroticism in West Indian Pentecostal community." *British Journal of Clinical Psychology* **20**(Pt 4): 295-296.

Wasson, R. G., S. Kramrisch, J. Ott and C. A. P. Ruck (1986). *Persephone's Quest: Entheogens and the Origins of Religion.* New Haven Conn, Yale University Press.

Watson, J. D. (1968). *The Double Helix.* New York, Atheneum.

Watson, P. (2006). *Ideas: A History from Fire to Freud.* London, Phoenix House Ltd.

Watters, D. E. (1975). "Siberian Shamanistic Traditions Among the Kham-Magars of Nepal." *Contributions to Nepalese Studies* **2**(1): 123-168.

Webster, R. and S. Holroyd (2000). "Prevalence of Psychotic Symptoms in Delirium." *Psychosomatics* **41**: 519-522.

Weiner, A. B. (1998). *The Trobrianders of Papua New Guinea.* Fort Worth, Texas, Rinehart and Winston.

Weisbrod, M., S. Maier, S. Harig, U. Himmerlsbach and M. Spitzer (1998). "Lateralised semantic and direct semantic priming effects in people with schizophrenia." *British Journal of Psychiatry* **172**: 142-146.

Weisfeld, G. E. (1993). "The Adaptive Value of Humor and Laughter." *Ethology and Sociobiology* **14**: 141-169.

Weiss, K. M. (1993/1995). *Genetic Variation and Human Disease: Principles and evolutionary approaches.* Cambridge, UK Cambridge University Press.

Weissman, D. H., A. S. Perkins and M. G. Woldorff (2008). "Cognitive control in social situations: A role for the dorsolateral prefrontal cortex." *Neuroimage* **40**(2): 955-962.

Wells, D. S. and D. Leventhal (1984). "Perceptual Grouping in Schizophrenia: Replication of Place and Gilmore." *Journal of Abnormal Psychology* **93**(2): 231-234.

White, T., M. Nelson and K. O. Lim (2008). "Diffusion tensor Imaging in Psychiatric Disorders." *Topics in Magnetic Resonance Imaging* **19**: 97-109.

Whitwell, F. D. and M. G. Barker (1980). " 'Possession' in psychiatric patients in Britain." *British Journal of Medical Psychology* **53**(4): 287-295.

Whitworth, J. M. (1975). *God's Blueprints: A Sociological Study of Three Utopian Sects.* London, Routledge & Kegan Paul Ltd.

Wiersma, D., J. A. Jenner, G. Van De Willige, M. Spakman and F. J. Nienhuis (2001). "Cognitive behaviour therapy with coping training for persistent auditory hallucinations in schizophrenia: a naturalistic follow-up study of the durability of effects." *Acta Psychiatrica Scandinavica* **103**: 393-399.

Williams, G. C. (1966). *Adaptation and Natural Selection*. Princeton, Princeton University Press.

Williams, H. J., M. J. Owen and M. C. O'Donovan (2009). "Schizophrenia genetics: new insights from new approaches." *British Medical Bulletin* **91**: 61-74.

Wilson, B. (2008) from http://www.brianwilson.com/brian/musicians.html, accessed 2010).

Wilson, D. R. (1993). "Evolutionary epidemiology: Darwinian theory in the service of medicine and psychiatry." *Acta Biotheoretica* **41**: 205-218.

Wilson, D. R. (1998). "Evolutionary epidemiology and manic depression." *British Journal of Medical Psychology* **71**: 375-395.

Wilson, D. S. (2002). *Darwin's Cathedral: Evolution, Religion and the Nature of Society*. Chicago, University of Chicago Press.

Wilson, D. S. (1997). "Introduction: multilevel selection theory comes of age." *American Naturalist* **150**(Supplement): 1-4.

Wilson, D. S. and L. Dugatkin (1997). "Group selection and Assortative Interactions" *The American Naturalist* **149**(2): 336-351.

Wilson, D. S. and E. Sober (1994). "Reintroducing group selection to the human behavioral sciences." *Behavioral and Brain Sciences* **17**: 585-654.

Wilson, E. O. (2012) *The Social Conquest of Earth*. New York, Liveright Publishing Corporation.

Wilson, E. O. (1975). *Sociobiology: The New Synthesis*. Cambridge, MA, Belknap Press.

Wilson, S. C. and T. X. Barber (1981). Vivid fantasy and hallucinatory abilities in the life histories of excellent hypnotic subjects. *Imagery: Concepts, Results and Applications*. E. Klinger. New York, Plenum.

Wilson, S. C. and T. X. Barber (1982). "The Fantasy-Prone Personality: Implications for understanding imagery, hypnosis and parapsychological phenomena". *Imagery: Current Theory, Research and Applications*. A. Sheikh. New York, John Wiley & Sons.

Winkelman, M. (1989). "A cross-cultural study of shamanistic healers." *Journal of Psychoactive Drugs* **21**(1): 17-24.

Winkelman, M. (1990). "Shamans and Other 'Magico-Religious' Healers: A Cross-Cultural Study of Their Origins, Nature, and Social Transformations." *Ethos* **18**(3): 308-352.

Winkelman, M. (2004). "Shamanism as the original neurotheology." *Zygon* **39**: 193-217.

Winterer, G. (2010). "Why do patients with schizophrenia smoke?" *Current Opinion in Psychiatry* **23**(2): 112-119.

Wolf, S. A., A. Melnik and G. Kempermann (2011). "Physical exercise increases adult neurogenesis and telomerase activity, and improves behavioral deficits in a mouse model of schizophrenia." *Brain, Behavior, and Immunity* **25**(5): 971-980.

Woodward, T. S., S. Moritz, M. Menon and R. Klinge (2008). "Belief Inflexibility in Schizophrenia." *Cognitive Neuropsychiatry* **13**(3): 267-277.

Wool, D. (2006). *The Driving Forces of Evolution: Genetic Processes in Populations.* Enfield, NH, Science Publishers.

Wrangham, R. and D. Peterson (1996). *Demonic Males: Apes and the Origins of Human Violence.* Boston, Mariner Books.

Wright, P. A. (1989). "The Nature of the Shamanistic State of Consciousness: A Review." *Journal of Psychoactive Drugs* **21**(1): 25-33.

Wright, R. (2009a). *The Evolution of God.* New York, Little Brown & Co.

Wright, R. (2009b). "Do Shamans Have More Sex? New Age spirituality is no more pure than old-time religion." *Slate.*

Wynne-Edwards, V. C. (1962). *Animal Dispersion in Relation to Social Behavior.* New York, Hafner Publishing Co.

Xiao, J., J. Li, L. Yuan and S. D. Tanksley (1995). "Dominance Is the Major Genetic Basis of Heterosis in Rice as Revealed by QTL Analysis Using Molecular Markers." *Genetics* **140**: 745-754.

Yap, P. M. (1960). "The Possession Syndrome: A comparison of Hong Kong and French findings." *The British Journal of Psychiatry* **106**: 114-137.

Yates, R. and A. Manhire (1991). "Shamanism and Rock Paintings: Aspects of the Use of Rock Art in the South-Western Cape, South Africa." *The South African Archaeological Bulletin* **46**(153): 3-11.

Yip, K. S. (2003). "Traditional Chinese religious beliefs and superstitions in delusions and hallucinations of Chinese schizophrenic patients." *International Journal of Social Psychiatry* **49**(2): 97-111.

Yolken, R. H. and E. F. Torrey (2008). "Are some cases of psychosis caused by microbial agents? A review of the evidence." *Molecular Psychiatry* **13**: 470-479.

Yucel, M., E. Bora, D. I. Lubman, N. Solowij, W. J. Brewer, S. M. Cotton, P. Conus, M. J. Takagi, A. Fornita, S. J. Wood, P. D. McGorry and C. Pantelis (2010). "The Impact of Cannabis Use on Cognitive Functioning in Patients With Schizophrenia: A Meta-analysis of Existing Findings and New Data in a First-Episode Sample." *Schizophrenia Bulletin.*

Zanarini, M. C., A. A. Williams, R. E. Lewis, R. B. Reich, S. C. Vera, M. F. Marino, A. Levin, L. Yong and F. R. Frankenburg (1997). "Reported pathological experiences associated with the development of borderline personality disorder." *American Journal of Psychiatry* **154**(8): 1101-1106.

Znamenski, A. A., Ed. (2004). *Shamanism: Critical Concepts in Sociology.* London, RoutledgeCurzon.

Znamenski, A. A. (2007). *The Beauty of the Primitive: Shamanism and the Western Imagination.* Oxford, Oxford University Press.

Zuardi, A. W., J. A. S. Crippa, J. E. C. Hallak, F. A. Moreira and F. S. Guimaraes (2006). "Cannabidiol, a Cannabis sativa constituent, as an antipsychotic drug." *Brazilian Journal of Medical and Biological Research* **39**: 421-429.

Index

Aborigines, Australian, 45, 46, 132, 139, 146
Ackerknecht, E. H., 153, 156
adoption studies, 30, 31
age of onset of schizophrenia, 36
Age of Reason, 43
alcohol use, 53, 77, 79, 89, 180, 214f
Alcorta, C., 143
alphabet, invention of, 10
Altamira, Cave of (Spain), 128
Alternate Uses Test, 85
altruism, 79, 110, 112–114, 116, 117–118, 133, 144, 148
Alzheimer, Alois, 21, 51
amanita muscaria (fly agaric), 178
American Civil War, 165
anal personality, 19
antagonistic pleiotropy, 107, 108
Applewhite, Marshall, 185
Aquinas, Thomas, 184
Aranda (Australia), 171
Ariely, Dan
  Predictably Irrational, 119
Aristotle, 7, 86
Artbeat Studio (Winnipeg), 210
Asahara, Shoko, 188
assortative mating, 116, 117, 214, 215, 219, 220
atheism, 101, 148
atlatl, 131
Atran, Scott, 141, 146
  In Gods We Trust: The Evolutionary Landscape of Religion, 141
attachment theory, 18
attention deficit disorder, 75
Augustine, Saint, 184
autism and autistic tendencies, 22, 67, 69, 70, 75, 80, 81, 86, 89, 91, 106, 220
autoimmune disorders, 64
ayahuasca, 179
Aymara people (Peru), 170
Aztecs, 146, 178

Bachrach, Bernard, 38
Bacon, Francis, 16
Baldwin effect, 221
barbiturates, 22
Barkow, Cosmides and Tooby
  The Adapted Mind, 5
Barrett, Syd (Pink Floyd), 91
Bart, Nigel, 210
Bayesian inferences, 81
Beatles, 188
Beattie, James, 16
Beck, Aaron, 24
Belsky, Jay, 75
Bena Bena people (New Guinea), 163
Berrios, G.E., 40
Bethlem Royal Hospital (London), 39, 41, 44, 92
Bhatia, T. et al., 33
Bhugra, Dinesh, 40, 72
Bible, 92, 146, 171, 182
bipolar disorder, 23, 25, 31, 39–41, 50, 53, 58, 61, 67, 69, 72, 75, 77, 84, 87, 89–92, 174, 181, 184–186, 203, 210, 213–215
Blake, William, 92
Blanchard, Kendall, 136
Bleuler, Eugen, 22, 37
  four As of schizophrenia, 22, 37
blindspot (optic disc), 65
Boas, Franz, 153
Bogoras, Waldemar
  see Bogoraz, Vladimir
Bogoraz, Vladimir, 154, 155
Bonferroni correction, 11
borderline personality disorder, 23, 77, 105, 120, 186
Bowlby, John, 18
Bowles, Samuel, 112
Boyer, Pascal, 135, 143, 146, 170–173, 184, 189, 211, 212
brain atrophy, 53
Brandon Asylum (Brandon Mental Health Centre), 203
Brewerton, 71
Brugger, Peter, 177

Brüne, Martin, 122, 211
  Textbook of Evolutionary Psychiatry, 122
Buddhism, 15
Burch, G. S. J. et al., 85
Burns, Jonathan, 108
Burton, Robert, 37

Caligula, 93
Campbell, Darren, 97
canalization, 103
cancer, 66
cannabis (marijuana), 26, 161, 179, 180, 181
Cardno, A. G., 30
Carmelite nuns, 141
Carpenter, Edmund, 164
categorical approach to schizophrenia, 33, 34
Celexa, 23
chandelier neuron axon terminals, 52
Charaka Samthita, 40
Charles VI (king), 93
Chauvet cave paintings (France), 126, 128
Cherokee people (North America), 136, 144
Chewong people (Malay Peninsula), 144
Chiarugi, Vincenzo, 20, 21
chlorpromazine, 23, 203
Chomsky, Noam, 101, 132
Christian Science, 184
Chukchee people (Russia), 154–156
Church of Jesus Christ of Latter-Day Saints, 184
Churchill, Winston, 8
classical conditioning, 17
classifications of mental disorders
  ancient Chinese, 40
  ancient Greek, 40, 42
  ancient India, 40, 41
  ancient Roman, 42
Clegg, Helen, 86, 107
Cleomenes I (king of Sparta), 42
clozapine, 23
cognitive behavioural therapy (CBT), 24, 162
cognitive deficits, 57, 58, 59
cognitive modules/modularity, 122, 146
cognitive-emotional states, 79, 80
complex communication, 113–115, 131
COMT (catechol-O-methyltransferase), 61
connectionism/connectionist model, 56–57, 87
Cooper et al., 68, 69
Copenhagen High Risk data set, 215
Copper Inuit (Canada), 130
copy number variation (CNV), 61, 62, 103, 104
Crick, Francis, 102
CT (computed topography) scan, 52, 53
cults, 169, 170, 182, 185, 186, 188, 189, 218, 219, 222
cyto-architectural abnormalities, 52

cytomegalovirus, 52

d'Aquili, E.G., 141
Da Costa, Jacob, 165
  Da Costa's syndrome, 165
Dakin, Steve et al, 96f
Dani people (New Guinea), 171
Darwin, Charles, 17, 25, 93, 99–102, 111, 112, 117, 181, 209
  On the Origin of the Species, 17, 102, 111, 117, 209
  The Descent of Man, 17
datura, 179
Dawkins, Richard
  The Selfish Gene, 104
Dawson, James, 138
Deacon, Terrence, 101
defense mechanisms, 19, 80
Dein, S., 211
déjà vu, 77, 142
Delay, Jean, 23
delirium, 16, 26, 39, 40, 41, 50, 70, 72–74, 77
delusional subtypes, 173
delusions, 25–27, 78, 211
dementia, 26, 41, 50, 66, 67, 70, 72, 74, 89, 93, 172, 192, 203
dementia praecox, 21, 204
Deniker, Pierre, 23
Dennett, Daniel
  Darwin's Dangerous Idea, 101
depression, clinical, 23–25, 40, 41, 67, 69, 74, 77, 79, 92, 93, 105, 117, 120–122, 183, 203, 207, 214, 215, 216
  learned helplessness, 120
Descartes, Rene
  philosophy of dualism, 17
Devereux, George
  Mohave Ethnopsychiatry and Suicide, 157
Di Mambro, Joseph, 185
Dianetics, 185
Dickens, Charles, 92
dimensional approach to schizophrenia, 33, 34, 83
directional evolution in transit, 106
DISC1 (disrupted-in-schizophrenia 1), 61
disease-phenotype hybrids, 64, 68
divination, 145, 166
  water witching, 145
DNA, 32, 45, 46, 49, 60, 102–105
DOAO (D-amino acid oxidase inhibitor), 61
Dobzhansky, Theodosius, 209
Dorobo people (Africa), 130
double bookkeeping phenomenon, 3
Down syndrome, 68
dream interpretation, 19
Driscoll, Carlos et al., 131
drug-induced psychosis, 26
DSM-III, 24, 30, 42

DSM-III-R, 30
DSM-IV, 39, 85, 192
DTI (diffusion tensor imaging), 52, 53
DTNBP1 (dysbindin), 61
Dunbar, Robin, 132
Durkheim, Émile, 144, 221
dwarfism, African Pygmies, 68
Dylan, Bob, 187

E-Prime, 97
Eastwell, H. D., 46
Ebbinghaus illusion (Titchener circles), 95, 96f
Eddy, Mary Baker, 184
Edison, Thomas, 181
EEG (electroencephalography), 59
Effexor, 23
ego, 19
Einstein, Albert, 4, 181
electro-convulsive therapy (ECT), 22, 72, 203
Eliade, Mircea, 163
Ellis, Thomas, 146
Endler, J. A., 109
Enlightenment philosophers, 16
entheogens, 178–180, 182
entoptic images, 127, 128
Epictetus, 16
epilepsy, 50, 69–71, 74, 158
    temporal lobe epilepsy (TLE), 26, 71, 77, 141
ESP (extra sensory perception), 176, 177
ethology, 18, 99, 117–121
eusocial insects, 113–115, 116f, 219
Evenki people (Siberia), 151, 154, 156, 161, 163
evil eye, 146
evolutionary disease model, 63–66
evolutionary tradeoff, 108
Exorcist, The, 175
experimental psychology, beginnings of, 16, 17
Eysenck Personality Questionnaire, 85

false beliefs, 79, 80, 82
Faraday, Michael, 93
fecundity
    of relatives of schizophrenia patients, 76, 83
    of schizophrenia patients, 29, 31–34, 46, 83, 105
Feyerbend, Paul, 12
Fingo tribe (South Africa), 157
Fischer, Bobby, 91
fixed action patterns, 18
Fleming, Alexander, 88
fMRI (functional magnetic resonance imaging), 52, 54, 55, 141, 142, 211
folk healers (Bali), 164
Fosbury, Dick
    Fosbury Flop, 90

Foucault, Michel, 43
Fox, George, 183
free association, 19
Freud, Sigmund, 18, 19, 63, 93, 181
Frith, Chris, 211
Fumane Cave (Italy), 127

Galileo, 12
genes + time = evolution, 4
genetic diseases, 64
genetic drift, 106
George III (king), 39, 93
Gilmore, G. C., 94, 95, 97
gliosis, 51
Goodall, Jane
    Through a Window, 117
Gould, Glenn, 91
gradualism, 100
Grim, John A., 168
group selection
    see multilevel (interdemic) selection
group splitting, 123, 124
Guilford's Alternate Uses Test, 93

Haizmann, Christoph, 42
HAL
    2001: A Space Odyessy, 170
hallucinations
    auditory, 25–27, 40, 77, 78, 91, 92, 154, 159, 162, 172, 183, 186
    tactile, 77
    visual, 26, 27, 72, 77, 78
Hardy-Weinberg equation, 31
Harlow, Harry, 120
healing, 143, 158, 160, 166, 167
Heaven's Gate, 185
Heinrichs, R.W., 57
Hemingway, Ernest, 92
Hendrix, Jimi, 187
Henry VI (king), 93
heritability, 29, 30, 31, 33, 45, 46, 60, 67, 69, 102, 107, 115, 116, 153, 209, 215, 216
Herodotus, 42
heterosis, 108, 109, 219
heterozygote advantage, 108–110, 116, 219, 220
Hippocrates, 16
Hitler, Adolf, 123, 186, 187
Hobbes, Thomas, 16, 39
Hoffer, Abram, 29
Holroyd, S., 72
Horne, David J. DeL, 45, 46
Hubbard, L. Ron, 185
Hume, David, 16
humor, 16, 58, 59, 119, 129, 132, 134, 136, 210, 212, 213, 218
Hunt, George (Tlingit shaman), 153
Hurley, R. A., 211

Huxley, Julian, 29, 122
Huxley, T. H., 100
hybrid vigor
  see heterosis
hypofrontality, 55
hypothesis testing
  falsification, 12

id, 19
Ignatius of Loyola, 184
incidence of schizophrenia, 35
industrialization, 43–45
insane asylums, 16, 20, 21, 44, 45, 203
insight-oriented psychotherapy, 23, 24
insulin coma treatment, 22, 72, 73, 203
inter-rater reliability, 2, 35
Iroquois people (North America), 152

James, William
  The Varieties of Religious Experiences, 140
Jamison, Kay, 92
Jeste, Dilip V. et al., 37, 41, 42
Jesuit missionary (North America), 152
Jesus, 183, 185
jimson weed
  see datura
Jivaroan shamans (Peru), 179
Joan of Arc, 186
Jones, I. H., 45, 46
Jones, Jim, 188
jongleurs (jugglers), 152
Jouret, Luc, 185
Joyce, James, 92
Jung, Carl, 187
Junod, Henri A., 158

Kabyle people (Algeria), 175
Kafka, Franz, 221
Kahlbaum, Karl, 63
Kallmann, Franz, 30
Kant, Emmanuel, 16
Karlsson, Jon
  Icelandic study, 84, 90
Kauffman, C. et al., 84
Kaxinawá people (Peru), 179
Keefe, Richard, 93
Kekulé, August, 88
Keller, Matthew, 106–108
kenaima, 138
Kendler, K.S., 62
Kety, Seymour, 30
Khaitovich, Philipp et al., 62
Kibweteere, Joseph, 185
kin selection, 110, 113
Kishimoto, M., 61

Kolob (planet), 184
Koresh, David, 188
Kraepelin, Emil, 21, 22, 30, 37, 40, 214
Kroll, Jerome, 3843
Kropotkin, Peter, 111
kuang, 40
Kuhn, Thomas
  The Structure of Scientific Revolutions, 7

La Barre, Weston, 166, 182
Lam, L. C. W., 40
laryngeal nerve, 65
Lascaux caves (France), 128
Laubscher, B. J. F.
  Sex, Custom and Psychopathology, 157, 158
Lauronen, Erika et al., 90
Leary, Timothy, 140
Lee, General Robert E., 133
Lee, Mother Ann, 183
Lee, Nathaniel, 92
Lennon, John, 187, 188
Levitin, Daniel, 133
Lewis, I. M., 161
Lindberg, David C.
  The Beginnings of Western Science, 11
Linde, Paul R., 162
Linnaeus, Carl, 88
Little Shop of Horrors, 170
Littlewood, R., 211
lobotomy, 22, 73, 203
Locke, John, 16, 39
Loeb, E. M., 158, 163
Lorenz, Konrad, 18
Ludwig II (king), 93
Luther, Martin, 184
Lyell, Charles
  Principles of Geology, 100

Maasai shamans (Africa), 145
Macaulay, Lord, 183
Machiavelli, Niccolo, 16
MacNeill, Allen D., 147
Macushi tribes (Guyana), 137–139, 167
magico-religious delusions, 193, 203, 207, 214
Magora, P.A., 93
Malur, C. et al., 72
maniacal eyes, 160
manic-depression
  See bipolar disorder
Manitoba Asylum (Selkirk Mental Health Centre), 203
Manson, Charles, 188
Maravich, Pete, 90, 91
Marx, Karl, 44, 181, 221
Mas, A. C., 41
materialism, philosophy of, 17
maternal separation anxiety, 121

Mayr, Ernst, 29, 122
Mazatec people (Mexico), 179
McCartney, Paul, 188
McGuire, Michael
  Darwinian Psychiatry, 122
Mendel, Gregor, 93, 102, 181
micropsia, 77
Middle Ages, 20, 38, 41, 43, 175
Miller, Geoffrey, 86, 106–108
Milton, John, 92
Mithen, Steven, 126
Mohammed, 183, 184
Mohave people (North America), 157
money, invention of, 10
Moniz, Egas, 73
monotheism, 135, 137, 139, 144, 146, 169, 170
mood-ridden psychosis
  See bipolar disorder
Moore, Omar, 145
Mormon, Book of, 184
Mormons, 184
Moroni (angel), 184
Moses, 183
Movement for the Restoration of the Ten Commandments of God, 184
MRI (magnetic resonance imaging), 52, 53
MRS (magnetic resonance spectroscopy), 52, 54
multilevel (interdemic) selection, 109–113, 116, 129, 191
multiple sclerosis, 69, 74
Munch, Edvard, 92
Murphy, Jane, 164
Murray, R. M., 30
mutation-selection balance, 106
Muuruup, 138
myelin dysfunction, 53

N-acetylaspartate (NAA), 54
Nash, John, 93, 94
Naskapi people (Labrador), 145
Nazism, 187
Neanderthals, 129, 131
negative frequency-dependent selection, 106, 107
Nelson (parrot), 118
Nepalese shamans, 158
Nesse, Randolph, 63, 64, 66–68, 108, 122
  Why We Get Sick, 63, 122
Nettle, Daniel, 85, 86, 107
Nettles, Bonnie, 185
neuregulin (NRG1), 54, 61
neuropsychiatric disorders
  chromosomal abnormalities, 68, 69
  polygenic diseases, 69, 74
  single gene defects, 69

neuroreceptors, 50
  alpha-7 nicotine, 51
  dopamine, 50, 61
  gamma-Aminobutyric acid (GABA), 51
  glutamate, 50, 51
  serotonin, 50, 142, 166, 219
Newberg, Andrew, 141
Newton, Isaac, 7, 12, 92, 181
nicotine/tobacco, 179, 180, 181
nicotinic acetylcholine receptors, 99
Nietzsche, Friedrich, 93
Nissl, Franz, 21
Niue, island of (Polynesia), 163
Noll, Richard, 153
Norman, Moe, 91
Nowak, Martin, 112
NRG1 (neuregulin 1), 61

obsessive compulsive disorder (OCD), 23, 43, 67, 69, 70, 73, 75, 105, 106, 121, 122, 165, 216
Oedipal complex, 18
Oesterreich, T. K., 175
ontological categories, 170–172, 212
operant conditioning, 18, 80
  Skinner box, 18
Order of the Solar Temple, 185
Osmond, Humphry, 29
Otsuka et al., 42
Owen, M. J. et al., 62, 94

Pahnke, Walter, 140, 179
paranoia and paranoid delusions, 26, 27, 64, 72, 76, 79, 81–83, 91–93, 158, 167, 174, 191–193, 203, 207, 218
Parnas, J., 215
Pavlov, Ivan
  Pavlovian responses, 17
Paxil, 23
peaiman (shaman), 138
penis envy, 19
Peters Delusional Inventory (PDI), 82
peyote, 179
Philo, Chris, 38, 42–44
phrenitis
  see delirium
physicalism, philosophy of, 17
Pinel, Philippe, 21
Pinocchio, 170
Pirnmeheeal, 138
Place, E. J. Schwartz, 94, 95, 97
Place-Gilmore experiment, 94, 95f
Plato, 16, 86
pleasure principle, 18
pneumoencephalography, 52, 53
polygenic mutation-selection balance theory, 106
polymorphism, 31, 33, 83, 105–109, 116

Polynesian shamans, 158
polyploidy, 103
polytheism, 11, 137–139, 143, 144, 146
Popper, Karl, 12
Post, 89
posttraumatic stress disorder, 43, 77, 165, 187, 207
Potts, S. G., 73
Pound, Ezra, 92
pre-pulse inhibition (PPI), 59
Price et al., 67
Price, John, 121–123, 182, 186
Prince (musician), 187
programmed death, 66, 67
Prozac
  fluoxetine, 23
psilocybin, 141, 142, 178, 179
psychological determinism, 19
psychotherapy/psychoanalysis, 16, 19, 20
Pythagoras, 92, 185, 186

Quakers
  see Religious Society of Friends

Regier, Darrel A., 25
Reiss, Jeff, 1, 6, 55, 97, 123, 132, 191
Religious Society of Friends, 183
RELN (reelin), 61
Renaissance, 16, 37, 41
repetition compulsion, 19
RGS4 (regulator of G-protein signaling 4), 61
Riel, Louis, 186
Riley B., 62

Samburu shamans (Africa), 145
Sasaki, Yuji, 159
Sawatsky, Breanna, 97
scala naturae, 101
scapulimancy, 145
schizotype (schizotypy, schizotypal), 3, 34, 85, 89, 107, 123, 124, 176, 182, 186–188, 220
  and creativity, 85, 107
Schneider, Kurt, 172, 173, 177
Schreber, Daniel, 93
Schuldberg, D. et al., 85
Scientology, Church of, 185
Scratching-Woman (shaman), 154, 155
Seligman's Error, 165
Seligman, Martin, 120
semantic memory, 87, 88
Seneca, 86
Seventh-day Adventist Church, 184
sex ratios of schizophrenia, 36
Shakers, 183
Shirokogoroff, S.M., 153, 155, 156

Psychomental Complex of the Tungus, 155, 156
Sieroszewski, Wacław, 155
Silverman, Julian, 2
single nucleotide polymorphism (SNP), 103, 104
Skinner, B. F., 17, 18
Smith, Joseph, 184
smooth-pursuit eye movement, 60
Sober, E., 112
social identity theory, 133
sociobiology, 119
sociopathy, 107
Socrates, 92
soldier's heart, 165
Sosis, Richard, 143
Spencer, Herbert, 100
Spinoza, Benedict, 16
spiritual experiences/delusions, 38, 71, 72, 80, 139, 141, 214
spirituality, 6, 8, 71, 76, 136–140, 137–142, 144, 147, 149, 153, 169, 170, 174, 189, 212
Spitzer, Robert, 24
Stephen, M., 164
Stevens, Anthony, 123, 182, 186
  Evolutionary Psychiatry, 122
stress diathesis model, 104, 220
suicide, 27, 90, 92, 93, 185, 203
sundowning, 26
super-ego, 19
superstitious thinking, 80
supportive psychotherapy, 23
supranormal stimuli, 18
Suryani, L. K., 164
Swedenborg, Emanuel, 93, 184
symptoms of mania
  DSM-IV, 39
symptoms of schizophrenia
  DSM-IV, 39
  Kurt Schneider, 172, 173, 177
  negative, 26, 37, 39
  positive, 26
Szasz, Thomas, 8

Taber, K. H., 211
Tapirapé people (Brazil), 162
Tarahumara people (Mexico), 136
task specialization, 113–115, 219, 220
Tembu tribe (South Africa), 157
temporal-spatial variation selection, 107
Tennyson, Alfred Lord, 92
Teresa of Avila, Saint, 181
Thales of Miletus, 9–11
Thomas, Keith, 37–39
Thonga people (South Africa), 153, 158
Thorndike, Edward, 17
Thurn, Everard, 137, 138, 167
  Among the Indians of Guiana, 137

Tinbergen, Nikolaas, 18
Tishkoff, Sarah, 64
Tlingit people (North America), 153
toxoplasma gondii, 52
trance states, 160, 161
transcendental experiences, 140, 141
transference, 19, 20
transposable genetic elements (transposons), 103
Trobrianders (New Guinea), 171
Troisi, Alfonso
  Darwinian Psychiatry, 122
Tungus people
  see Evenki people
twin studies, 30, 57, 142

Uhlhaas, P.J. et al., 95
Urubu people (Brazil), 171

Van Crevald, Martin, 130
van Gogh, Vincent, 92
van Os, Jim, 76
variable number of tandem repeats (VNTR), 103
Veatch, Thomas, 132, 212
Veblen, Thorstein
  The Theory of the Leisure Class, 119
vestigial traits/behaviors, 110, 136
Vodou (Haiti), 135, 171
von Frisch, Karl, 18

Waddell, C., 90
Wagley, Charles, 162
Watson, James, 102
Weber, Max, 221
Webster, R., 72
White, Ellen, 184

Williams syndrome (Williams-Beuren syndrome), 69
Williams, G.C., 112
Williams, George
  Why We Get Sick, 63, 122
Williams, H. J. et al., 62
Wilson, Brian (musician), 91
Wilson, D. S., 112
Wilson, Daniel R., 31
Wilson, David Sloan, 144, 148
Winkelman, Michael, 153, 154
Wisconsin Card Sorting Test, 58
Woolf, Virginia, 92
word association task, 55, 56, 88
World Health Organization Ten Country Study, 35
Wright, Robert, 137, 144, 146
  The Evolution of God, 33, 137
Wundt, Wilhelm, 17, 21
Wynne-Edwards, V. C.
  Animal Dispersion in Relation to Social Behaviour, 111, 112

Xingu tribes (Brazil), 171

Yakuts (Russia), 154, 155
Yanomamo (Venezuela), 159, 161, 167
Yellow Emperor's Classic of Internal Medicine, The, 40
Yupik community (Alaska), 164

Zakzanis, K. K., 57
Zimbabwe, University of, 162
ZNF804A (zinc finger protein 804A), 61, 62
Zoloft, 23
Zuev, Vasilii, 1

Colophon

Shamans Among Us was typeset using LaTeX. The text font is TeXGyre Pagella, design by the GUST e-foundry and inspired by Palatino, and the heading font is PT Sans Narrow by ParaType. The cover fonts are Palatino, designed by Hermann Zapf, and vincHand by JoeBob Graphics (`http://www.joebob.nl`).

Made in the USA
Las Vegas, NV
31 January 2024